Vol. 33. **Masking and Demasking of Chemical Reactions.** By D. D. Perrin
Vol. 34. **Neutron Activation Analysis.** By D. De Soete, R. Gijbels, and J. Hoste
Vol. 35. **Laser Raman Spectroscopy.** By Marvin C. Tobin.
Vol. 36. **Emission Spectrochemical Analysis.** By Morris Slavin
Vol. 37. **Analytical Chemistry of Phosphorous Compounds.** Edited by M. Halmann
Vol. 38. **Luminescence Spectrometry in Analytical Chemistry.** By J. D. Winefordner, S. G. Schulman, and T. C. O'Haver
Vol. 39. **Activation Analysis with Neutron Generators.** By Sam S. Nargolwalla and Edwin P. Przybylowicz
Vol. 40. **Determination of Gaseous Elements in Metals.** Edited by Lynn L. Lewis, Laben M. Melnick, and Ben D. Holt
Vol. 41. **Analysis of Silicones.** Edited by A. Lee Smith
Vol. 42. **Foundations of Ultracentrifugal Analysis.** By H. Fujita
Vol. 43. **Chemical Infrared Fourier Transform Spectroscopy.** By Peter R. Griffiths
Vol. 44. **Microscale Manipulations in Chemistry.** By T. S. Ma and V. Horak
Vol. 45. **Thermometric Titrations.** By J. Barthel
Vol. 46. **Trace Analysis: Spectroscopic Methods for Elements.** Edited by J. D. Winefordner
Vol. 47. **Contamination Control in Trace Element Analysis.** By Morris Zief and James W. Mitchell
Vol. 48. **Analytical Applications of NMR.** By D. E. Leyden and R. H. Cox
Vol. 49. **Measurement of Dissolved Oxygen.** By Michael L. Hitchman
Vol. 50. **Analytical Laser Spectroscopy.** Edited by Nicolo Omenetto
Vol. 51. **Trace Element Analysis of Geological Materials.** By R. D. Reeves and R. R. Brooks
Vol. 52. **Chemical Analysis by Microwave Rotational Spectroscopy.** By Ravi Varma and Lawrence W. Hrubesh

Chemical Analysis
by Microwave Rotational Spectroscopy

CHEMICAL ANALYSIS

A SERIES OF MONOGRAPHS ON
ANALYTICAL CHEMISTRY AND ITS APPLICATIONS

Editors

P. J. ELVING   J. D. WINEFORDNER

Editor Emeritus: **I. M. KOLTHOFF**

Advisory Board

Fred W. Billmeyer. Jr.
Eli Grushka
Barry L. Karger
Viliam Krivan

Victor G. Mossotti
A. Lee Smith
Bernard Tremillon
T. S. West

VOLUME 52

A WILEY-INTERSCIENCE PUBLICATION

JOHN WILEY & SONS
New York / Chichester / Brisbane / Toronto

Chemical Analysis by Microwave Rotational Spectroscopy

RAVI VARMA
Chemical Engineering Division
Argonne National Laboratory
Argonne, Illinois

LAWRENCE W. HRUBESH
Chemistry and Material Science Department
Lawrence Livermore Laboratory
Livermore, California

A WILEY-INTERSCIENCE PUBLICATION

JOHN WILEY & SONS
New York / Chichester / Brisbane / Toronto

Copyright © 1979 by John Wiley & Sons, Inc.

All rights reserved. Published simultaneously in Canada.

Reproduction or translation of any part of this work beyond that permitted by Sections 107 or 108 of the 1976 United States Copyright Act without the permission of the copyright owner is unlawful. Requests for permission or further information should be addressed to the Permissions Department, John Wiley & Sons, Inc.

Library of Congress Cataloging in Publication Data

Varma, Ravi, 1936–
  Chemical analysis by microwave rotational spectroscopy.

  (Chemical analysis ; v. 52)
  "A Wiley-Interscience publication."
  Includes index.
  1. Microwave spectroscopy. I. Hrubesh, Lawrence W., joint author. II. Title. II. Series.
  QD96.M5V37   543'.085   78-17415
  ISBN 0-471-03916-0

Printed in the United States of America

10 9 8 7 6 5 4 3 2 1

PREFACE

It is somewhat surprising that an analytical technique as powerful as microwave rotational spectroscopy has not been included in the major course of study for analytical chemists. Certainly, it has limitations, as do all other analytical tools, but the technique is every analytical spectroscopist's dream model for handling ultra-high resolution; yet it has not gained popularity. Still, its analytical utility was recognized early in its development (mostly by physical chemists who were using its inherent high resolution for basic molecular studies), and examples of specific analytical applications can be cited from the period of the 1940s and 1950s. The 1960s brought the development of microwave rotational spectroscopy instruments, but none of these were developed primarily for the analytical market. However, during this time, a few individuals who recognized the great potential for analytical application of microwave rotational spectroscopy did a considerable amount of work toward its analytical development. Currently, it remains a little known and little used analytical technique, but the authors feel that this situation is likely to change. Two main reasons for its expected revitalization are: the feasibility of simple, low cost, special purpose instruments; and the rise of industrial and energy-related applications for selective gas (vapor) monitors.

With a view toward the eventual popular analytical application of microwave rotational spectroscopy, the authors believe this time to be most appropriate for a book which addresses its analytical features. This book is written mainly for the potential analytical users of microwave spectroscopy. Those readers familiar with other spectroscopic techniques will find it easy to read, and they will be most likely to identify its outstanding analytical characteristics.

The authors believe that this book represents the current art in the analytical application of microwave rotational spectroscopy. It presents the readers with the basic theory of the technique and the properties which qualify its analytical utility. The book includes representative applications and provides a prospectus for future use. In addition, it includes a comprehensive list of references related to the analytical use of the technique.

The authors wish to acknowledge several individuals who have been most influential in the development of analytical microwave rotation spectroscopy. Most notable among these are Prof. E. Bright Wilson, Jr., Dr. Howard Harrington, Prof. Robert F. Curl, Dr. Leroy Scharpen, Mr. William White, and Prof. Edgar A. Rinehart. An acknowledgment also goes to Mr. Jack Frazer, whose recognition of the potential of microwave rotational spectroscopy for chemical analysis, has led to the development and extensive use of the technique at a major national laboratory (Lawrence Livermore Laboratory).

Ravi Varma wishes to express his deep appreciation to his former teacher, Professor E. Bright Wilson, Jr. without whose encouragement this book would not have been possible. This author's greatest and very personal gratitude, however, is due to his wife Mamata and his daughter Madhulika. Without their understanding and their patient attitude, it would not have been possible to undertake this major assignment.

Livermore, California, USA   Ravi Varma
Hinsdale, Illinois, USA   L. W. Hrubesh
April 1978

CONTENTS

1. INTRODUCTION ... 1

    1.1 Microwave Rotational Spectroscopy: Introduction ... 1
    1.2 Principles of Microwave Rotational Spectroscopy ... 3
    1.3 Techniques of Microwave Rotational Spectroscopy ... 18
    1.4 Intensities and Shapes of Rotational Transitions ... 23
    1.5 Absorption Coefficient, α_0, and the Integrated Absorption, α_{int} ... 27

2. INSTRUMENTATION ... 35

    2.1 Commercial Spectrometers ... 35
      a. Hewlett-Packard Model 8460A, Microwave Rotational Resonance Spectrometer ... 35
      b. Cambridge Scientific Instruments Analytical Microwave Spectrometer ... 39
    2.2 Millimeter Wave Spectrometers ... 42
      a. Video-type Millimeter Wave Spectrometer ... 42
      b. Millimeter Wave Fabry-Perot Spectrometer ... 43
    2.3 Computer Interfacing of Microwave Spectrometers ... 44
    2.4 Spectrometer and Absorption Cells for Detection and Monitoring of Chemical Species of Short Life ... 51
    2.5 Solid-state Diode Microwave Sources and Microwave Resonant Cavities for High-sensitivity Spectrometers ... 55
    2.6 High-temperature Microwave Spectrometer ... 58

CONTENTS

	Page
3. QUALITATIVE ANALYSIS	59
3.1 Resolution and Sensitivity of Microwave Spectra	59
3.2 Specificity in High-resolution Microwave Rotational Spectroscopy for Qualitative Analysis	62
3.3 Applications of High-resolution Microwave Spectroscopy in Indentification and Characterization of Molecular Species, Reaction Intermediates, and Free Radicals	68
3.4 Qualitative Analysis with Computer-controlled Microwave Spectrometers	74
3.5 Low-resolution Microwave-band Spectroscopy for Qualitative Analysis	77
3.6 Applications of LMWBS in Identification and Characterization of Relatively Large Organic Molecules	82
4. QUANTITATIVE ANALYSIS BY MICROWAVE SPECTROSCOPY	86
4.1 Peak Intensity of Rotational Transitions	87
4.2 Integrated Intensity Coefficients of Rotational Transitions	91
4.3 Techniques of Measurements of Peak Intensities and Integrated Intensities with Illustrative Applications	92
a. Direct Comparison Method	92
b. Use of PIN-diode Signal Calibrators	99
4.4 Techniques of Relative and Absolute Intensity Measurements	100
a. Bridge Method	104
b. Peak Height and HWHM, $\Delta\nu$, Determination of Relative Intensities	105
c. Measurement of Integrated Intensities	106
d. Absolute Intensity Measurements	106

		Page
4.5	Γ-Intensity Coefficients for Chemical Analysis under Conditions of Power Saturation	114
	a. Use of Calibration Procedures for Measurement of Γ-Coefficient	115
	b. Use of PIN Diode and Calibration Arm for Measurement of Γ_{max} Intensity	116
4.6	Sources of Uncertainty in Analytical Determination by Microwave Spectroscopy	120
	a. Relative Intensity Measurements	120
	b. Absolute Intensity Measurements	121
4.7	Sensitivity, Accuracy, and Precision of Quantitative Analysis by Microwave Rotational Spectroscopy	121
4.8	Type and Size of Sample and Sampling Considerations for Analysis by Microwave Rotational Spectroscopy	123

5. APPLICATIONS AND FUTURE POTENTIAL

5.1	Reaction Studies by Microwave Rotational Spectroscopy	126
5.2	Pyrolysis and Ozonolysis Studies	130
5.3	Isotope-ratio Measurements	133
5.4	Mixture Analysis by Microwave Rotational Spectroscopy	135
5.5	Air-pollution Measurements	145
5.6	Other Applications	150
	a. Dectectors for Gases in/from Solutions	150
	b. Engine-exhaust Analysis	151
	c. Cigarette-smoke Analysis	152
5.7	Inexpensive Analytical Spectrometers	154
5.8	Process-control Sensors	159
5.9	Portable Gas Monitors	170
5.10	Combined Gas Chromatography/Microwave Spectroscopy	180

REFERENCES	186
AUTHOR INDEX	201
SUBJECT INDEX	203

Chemical Analysis
by Microwave Rotational
Spectroscopy

CHAPTER

1

INTRODUCTION

1-1. MICROWAVE ROTATIONAL SPECTROSCOPY

The term "microwave rotational spectroscopy" (MRS) generally implies the study of absorption spectra associated with rotational transitions of molecules or molecular fragments (e.g., free radicals and ionic species) in the vapor phase at low pressures. The microwave range covers the electromagnetic radiation of wavelengths from 0.3 mm to approximately 30 cm (or frequencies of 1 to 1000 GHz). However, the range of wavelengths between 7 mm and 4 cm (or frequencies of 8 to 40 GHz) is most commonly used for analytical applications. The extent of the microwave region is shown in Fig. 1-1.

The first high-resolution microwave spectra [1-3] in the gas phase were observed for the ammonia molecule in 1946. Electronic devices and techniques for the generation and detection of microwaves were developed during the early 1940s in connection with the war efforts. This led to a rapid exploitation of the microwave region of the electromagnetic spectrum to yield information pertaining to molecular parameters of interest to chemists and physicists. Gas-phase microwave spectroscopy is now about 30 years old. Until a few years ago, the field had been predominated by efforts aimed at the determination of fundamental molecular properties such as structure, electric dipole moments, magnetic moments, nuclear quadrupole coupling constants, conformation, and energy differences of rotational isomers.

A change of emphasis began a few years ago. Researchers have been striving to find new applications. As a result, a considerable amount of work relating to qualitative and quantitative chemical analysis has been published during the last few years. This includes work relating to detection, characterization, and quantitative determination of a component

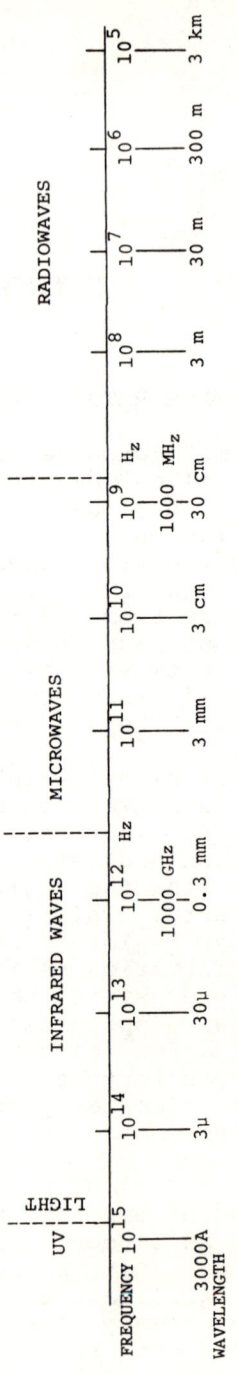

Fig. 1-1. Extent of the microwave region in electromagnetic spectrum.

or components in gas mixtures such as polluted samples of air or products from pyrolysis. Microwave rotational spectroscopy has been utilized to study the thermodynamic properties of pure substances and has recently found application in the study of chemical reactions. Rapid development of microwave electronics, automated microwave spectrometers, and computer interfacing with the spectrometers has led to renewed interest in applications in the field of chemical analysis.

This monograph is not intended as a complete survey of microwave rotational spectroscopic research. Rather, the aspects of microwave spectroscopy that bear on analytical applications are discussed in detail. The theory and experimental techniques of MRS are surveyed briefly to lay a proper foundation for a comprehensive understanding of the applications. The current practices of microwave rotational spectroscopy in terms of both theory and experimental technique, have been adequately discussed in excellent books [4-6] and review articles [7-13]. The analytical aspects of microwave spectroscopy have been discussed recently in general terms by Laurie [14] and Lovas [15].

1-2. PRINCIPLES OF MICROWAVE ROTATIONAL SPECTROSCOPY

According to the postulates of quantum theory, a molecular system absorbs or emits energy in discrete steps. One measures relative positions of two or more rotational energy levels by determining the frequencies of microwave radiation absorbed or emitted as a result of the interaction of the electric or magnetic components of the field of the microwave radiation with the electric or magnetic dipoles fixed in the rotating molecules. The microwave spectra of molecules consist of rotational transitions with various fine-structure effects, such as those due to the presence of nuclei with electric quadrupole moments in the molecule, superimposed. Polar molecules at low pressures will undergo rotational transition on exposure to monochromatic microwave radiation.

The frequency ν_{12}, of the transition between two rotational levels of energy E_1 and E_2 is obtained by applying the familiar Bohr postulate,

$$\nu_{12} = \frac{E_2 - E_1}{h} \tag{1-1}$$

where h = Planck's constant. The nature of the transition observed with a consequence of absorption or emission of the photon of energy, $h\nu_{12}$, depends to a large extent on the numbers of molecules N_1 and N_2 occupying respectively the lower and upper rotational states. Four different situations that develop as a direct consequence of relative population of molecules in the two rotational states are illustrated in Fig. 1-2. The types of rotational transitions resulting depend on the relative magnitude of N_1 and N_2 and are indicated by thick arrows. Case I (see Fig. 1-2) represents the Boltzmann distribution of population states. This is typical of the microwave spectral transitions observed for gases or vapors in waveguides at low pressures (>1 but <10 Pa) and low microwave power. It is possible to achieve a condition known as saturation (see Case II, Fig. 1-2) by increasing the radiation density (i.e., the microwave power level) to the extent that the rate of radiation-induced transitions is equal to the rate of transitions induced by molecular collisions. It is usually desirable that the radiation power level be kept low in an analytical determination to avoid the onset of saturation. However, if necessary, techniques are available for dealing effectively with problems due to saturation.

The less common situations described by Cases III and IV are usually achieved only by the use of special experimental arrangements. Narrow beams of molecules in rotational states E_2 or E_1 may be isolated by using molecular-beam techniques in conjunction with an appropriate electrostatic separator. For example, a MASER action (see Fig. 1-2, Case III) is observed by exposing a molecular beam in state E_2 to microwave radiation of ν_{12} frequency. Selecting a molecular beam in state E_1 under similar conditions would result in stimulated absorption shown in Fig. 1-2, Case IV.

Analytical applications of microwave spectroscopy require measurement of frequencies and intensities of transitions that usually correspond either to Case I or to the situation where both Cases I and II contribute.

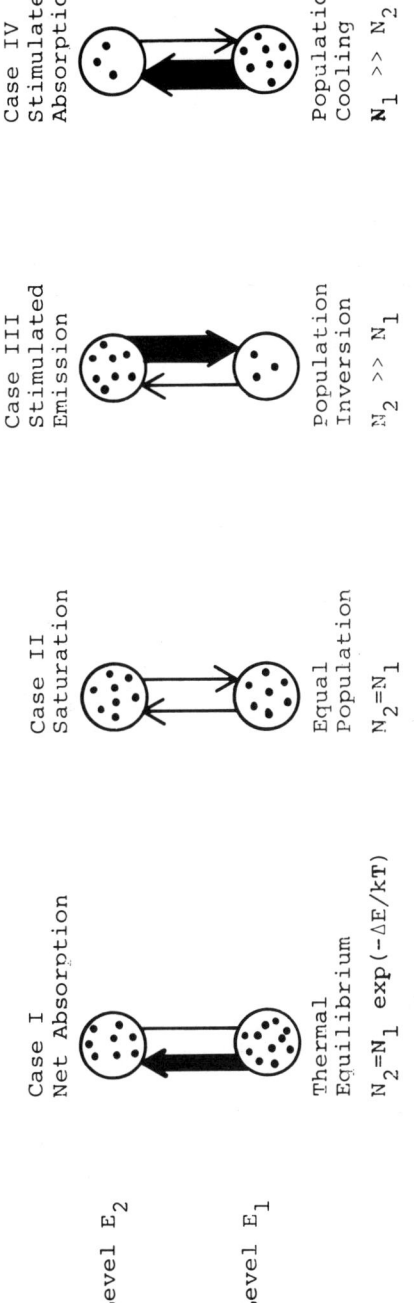

Fig. 1-2. Spectral consequences of difference in population of two rotational states.

The rotational motion of molecules can be described in terms of principal moments of inertia I_x, I_y, and I_z about the three body-fixed principal axes x, y, and z, respectively (see Fig. 1-3). Conventionally, principal axes are labeled a, b, and c with labeling determined by the relative sizes of the principal moments of inertia such that $I_a < I_b < I_c$. Rotational constants expressed in terms of the moments of inertia,

$$A = \frac{h}{8\pi^2 I_a}, \quad B = \frac{h}{8\pi^2 I_b} \text{ and } C = \frac{h}{8\pi^2 I_c}, \quad (1-2)$$

constitute another set of useful molecular parameters. Molecules are classified into a number of categories based on the relative sizes of the principal moments of inertia. The different classes along with some examples are given in Table 1-1.

A set of principal moments of inertia of a molecule can readily be calculated from a set of assumed or known structural parameters such as bond lengths and angles, as well as the atomic masses. The calculations are usually conducted in two stages. From the assumed structural parameters of the molecules, a set of Cartesian coordinates for all the atoms with reference to a chosen set of coordinate axes are obtained [16,17]. The principal moments of inertia

TABLE 1-1. TYPES OF ROTORS

Rotor	Principal Moments of Inertia	Example
Linear	$I_a (\sim 0) < I_b < I_c$	$HC\ell$, OCS
Spherical	$I_a = I_b = I_c$	SiH_4, CH_4
Prolate symmetric	$I_a < I_b = I_c$	SiH_3, GeH_3, POF_3
Oblate symmetric	$I_a = I_b < I_c$	BF_3, C_6H_6
Prolate asymmetric		SiH_3SiH_2F
Oblate asymmetric		Bf_2OH

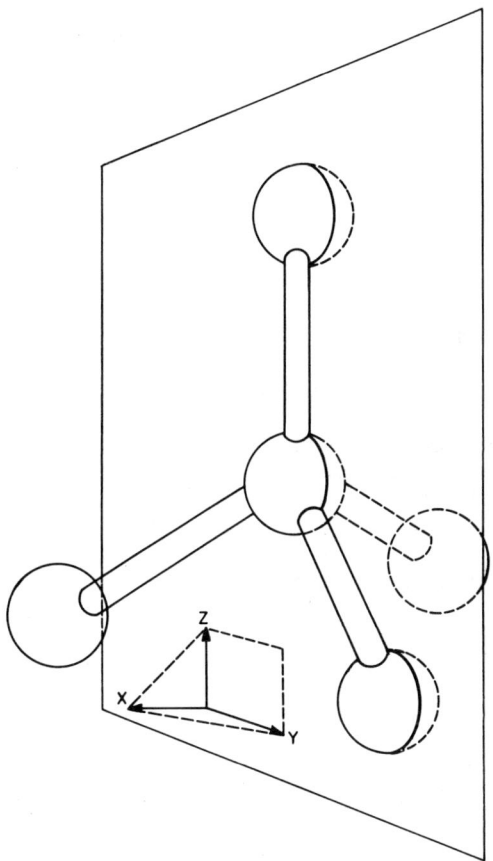

Fig. 1-3. Orientation of body fixed principal axes x, y, and z for a symmetric top molecule.

and the orientation of the inertia axes in the molecular frame may then be calculated from the Cartesian coordinates and masses of atoms as discussed in detail in the literature [4-6]. The rotational spectrum of any molecule may then be calculated from these estimated rotational constants. The absorption strengths for each transition requires knowledge of an electric dipole moment that may be estimated or obtained from Stark effect measurements. An experimentally observed spectrum may also be fitted to a set of rotational constants (A, B, and C). Such calculations of spectral lines from rotational constants, and vice versa, are extremely important for complete identification or characterization of a molecule.

The rotational energy levels of a linear and symmetric top rotor are adequately labeled by certain quantum numbers from which the energy may be calculated. The quantum numbers J, K, and M represent respectively the: (1) total angular momentum, $\hat{P}$, of the molecule, (2) component of $\hat{P}$ along the symmetry axis (z), $\hat{P}_z$, and (3) the component of $\hat{P}$ along the space fixed axis (Z), P_Z (see Fig. 1-4) in such rotors. The quantum number J can assume integral values from 0 to ∞. For a symmetric top, K can have values given by

$$K = J, J - 1, J - 2, \ldots, 0, \ldots, -J. \qquad (1-3)$$

For a linear molecule, with K as zero, J is used to label the levels. In the case of an asymmetric rotor the quantum number K is not a "good" quantum number. Instead a more complex labeling scheme must be used to specify the energy levels of an asymmetric rotor, as is discussed later in the text.

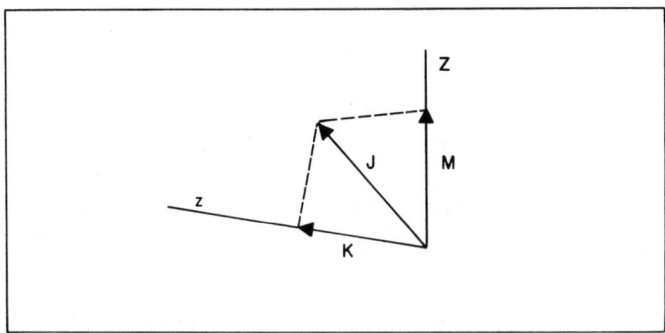

Fig. 1-4. Diagram showing the angular moment vectors; $\hat{P}$ (total angular momentum), $\hat{P}_z$ (component of P along the molecule fixed axis, z), and $\hat{P}_Z$ (component of $\hat{P}$ along a space fixed direction, Z), as well as their representation by J, K, and M quantum numbers.

The assumption that the rotational, vibrational, and electronic energies of a vibrating-rotating molecule in the gas phase are completely separable is a good approximation and is a necessary prelude to any calculation. Furthermore, molecules are considered to be rigid rotors in the first approximation. The Hamiltonian operator for any rigid rotor is given by the equation

$$\hat{H}_R = \frac{4\pi^2}{h} (A\hat{P}_a^2 + B\hat{P}_b^2 + C\hat{P}_c^2) \qquad (1\text{-}4)$$

where $\hat{P}_a$, $\hat{P}_b$ and $\hat{P}_c$ are angular momentum operators corresponding to the components of the total angular momentum, $\hat{P}$, of the top along principal axes a, b, and c, given by equation

$$\hat{P}^2 = \hat{P}_a^2 + \hat{P}_b^2 + \hat{P}_c^2 \qquad (1\text{-}5)$$

The angular momentum operators can be expressed as matrices [18] whose rows and columns are labeled by J and by K or M. The matrices representing the angular momentum operators $\hat{P}^2$, $\hat{P}_x$, $\hat{P}_y$, $\hat{P}_z$ or any algebraic functions of them can be generated very simply. One can write a matrix for the first term on the right-hand side of Eq. (1-4) by taking the matrix for P_a, multiplying it by itself, then scalar multiplying by A. This can similarly be done for the terms involving matrices $\hat{P}_b$ and $\hat{P}_c$, with each added to the first matrix. Thus the operator $\hat{H}_R$ [Eq. (1-4)] can itself be formed as a matrix. The energy levels of the rigid rotor are then found from the eigenvalues of the $\hat{H}_R$ matrix. The matrix elements of the Hamiltonian and the energy of a rotational state of a linear molecule ($I_z = I_a \approx 0$; $I_x = I_y = I_b$) are given by the equation

$$\langle JKM|\hat{H}_R|JKM\rangle = \frac{h^2}{8\pi^2 I_b} J(J+1) \qquad (1\text{-}6)$$

and

$$E = hBJ(J+1). \qquad (1\text{-}7)$$

Application of the selection rules, $\Delta J = \pm 1$, $\Delta K = 0$, and $\Delta M = 0, \pm 1$ for rotational transitions of linear molecules, yields the expression for frequencies in MHz,

$$\nu = 2B(J + 1), \quad (1-8)$$

where J is the quantum number designating the lower of any two states involved in the transition and B is the rotational constant expressed in MHz units.

The matrix elements of the Hamiltonian for prolate ($I_x = I_a < I_y = I_b = I_z = I_c$) and oblate $I_x = I_a = I_y = I_b < I_z = I_c$) symmetric rotors are given respectively by the equations

$$<JKM|\hat{H}_R|JKM> = \frac{h^2}{8\pi^2 I_b} J(J+1) + \frac{h^2}{8\pi^2}(\frac{1}{I_a} - \frac{1}{I_b}) K^2,$$

and $\quad (1-9)$

$$<JKM|\hat{H}_R|JKM> = \frac{h^2}{8\pi^2 I_b} J(J+1) + \frac{h^2}{8\pi^2}(\frac{1}{I_c} - \frac{1}{I_b}) K^2.$$

$$(1-10)$$

The corresponding energies for the rotational states are given by the equations,

$$E = hBJ(J+1) + h(A-B)K^2, \quad (1-11)$$

and

$$E = hBJ(J+1) + h(C-B)K^2. \quad (1-12)$$

For any value of J, K can have (2J + 1) values. Because energy is independent of the sign of K, levels with +K and -K coincide and are doubly degenerate. The level for which K = 0 is nondegenerate. Hence only (J + 1) rotational energy levels are possible for any value of J. Because of symmetry, the component of the dipole moment perpendicular to the unique axis of a symmetric top molecule is zero. Therefore, there is no torque in the direction of $\hat{K}$. Hence the angular momentum along the molecular axis does not change as a result of interaction with the microwave electric field. Therefore, the selection rules for J and K are $\Delta J = \pm 1$, $\Delta K = 0$. Use of these selection rules yields the expression for the frequencies of symmetric top rotor transitions,

$$\nu = 2B(J+1), \quad (1-13)$$

where J again refers to the lower rotational state. Thus the rotational spectrum of a linear as well as symmetric top rotor consists of lines at regular intervals separated from its closest neighbors by a frequency given by 2B.

All three principal moments of inertia of an asymmetric rotor are unequal to each other. The asymmetry of a molecule is most conveniently expressed in terms of an asymmetry parameter, κ, Ray's parameter [19] which is defined by,

$$\kappa = \frac{(2B - A - C)}{A - C} , \qquad (1-14)$$

and which can assume values in the range -1 (prolate symmetric rotor, $A > B = C$) to +1 (oblate symmetric rotor, $A = B > C$). There are $(2J + 1)$ energy levels for each value of J for an asymmetric top. These energy levels become degenerate K-states in the limiting symmetric rotor case. For an asymmetric rotor the total angular momentum, $\hat{P}$ (represented by the J quantum number), and its projection, $\hat{P}_Z$ (represented by the M quantum number), on the space-fixed direction defined by the direction of the applied Stark electric field in the absorption cell of a microwave spectrometer are constants of motion. Hence J and M are "good" quantum numbers. However, the quantum number K, representing the component of $\hat{P}$ along the appropriate molecular axis in symmetric rotor limits, loses its meaning in the case of an asymmetric molecule and is no longer a "good" quantum number. In the scheme for labeling the rotational energy levels in the case of an asymmetric rotor, the correlation with the near prolate and oblate symmetric rotor limits is emphasized. The levels can be labeled by J, K_-, and K_+ conveniently written in the form $J_{K_-K_+}$, where K_- and K_+ are the connecting quantum levels in the prolate and oblate symmetric rotor limits.

The asymmetric rotor Hamiltonian can be written in the form

$$\hat{H} = \frac{A + C}{2} (\hat{P}_a^2 + \hat{P}_b^2 + \hat{P}_c^2)$$
$$+ \frac{A - C}{2} (\hat{P}_a^2 + \kappa \hat{P}_b^2 - \hat{P}_c^2) \qquad (1-15)$$

where κ is Ray's asymmetry parameter. The energy levels of the rotor are the eigenvalues of the $\hat{H}$ matrix and may be written as

$$E_{JK_-K_+} = h\frac{A+C}{2}J(J+1) + h\frac{A-C}{2} \cdot \varepsilon_{JK_-K_+}(\kappa), \quad (1-16)$$

where $\varepsilon_{JK_-K_+}(\kappa)$ are the eigenvalues of the operator written as $\hat{H}_\kappa$ ($\hat{H}_\kappa = \hat{P}_a^2 + \kappa\hat{P}_b^2 - \hat{P}_c^2$). The energy levels can be obtained from the solutions of the secular equation

$$|\varepsilon_{JK_-K_+}(\kappa) - I\lambda| = 0 \quad (1-17)$$

derived from the reduced energy matrix corresponding to the reduced Hamiltonian, $\hat{H}_\kappa$. The problem of finding numerical solutions becomes simplified if one takes advantage of the symmetry properties of the asymmetric rotor Hamiltonian. The symmetry properties may be derived from the ellipsoid of inertia that is symmetric to an identity operation, E, as well as to C_2 rotations about its principal axes. The set of symmetry operations (E, C_2^a, C_2^b, and C_2^c) constitute what is known as the "four-group designated by V(a, b,c)." The rigid rotor Hamiltonian, $\hat{H}$ (including the one representing the asymmetric rotor), is invariant under the symmetry operations of this group. A knowledge of the symmetry properties of the Hamiltonian enables one to classify the quantum rotational states into one of the categories of the symmetry species A, B_a, B_b, and B_c. The notation, A, indicates invariance of the rotational wave function under all symmetry operations. The subscript of any of the other symmetry species (e.g., of B_a) indicates the particular principal axis of the molecule about which the wave function is symmetric or has a character of (+1). It is understood also that the function is antisymmetric about the other two axes, b and c, and has a character of (-1). This results in considerable simplification in the form of the energy matrix for an asymmetric rotor. Each asymmetric rotor wave function, $A_{JK_-K_+M'}$ can now be expressed as a linear combination of symmetrized Wang functions [20], $S_{JKM\gamma}$.

$$A_{JK_-K_+M'} = \sum_K a_K^{JK_-K_+M'} S_{JKM\gamma}. \quad (1-18)$$

The Wang functions are linear combinations of the symmetric rotor wave functions. The change from an expansion of the ψ_{JKM} (the symmetric rotor wave function) to an expansion of $S_{JKM\gamma}$ enables splitting of the energy matrix $\varepsilon_{J\ K_-K_+}(\kappa)$ into two submatrices corresponding to $\gamma = 0$ (symmetric Wang functions) and $\gamma = 1$ (antisymmetric Wang functions). Each of these matrices can be further factored into two matrices involving terms with even and odd K numbers. Each of these four submatrices may now be diagonalized separately and the reduced energy, $\varepsilon(\kappa)_{JK_-K_+M}$, obtained. The selection rules for J for an asymmetric rotor: $\Delta J = 0$ (Q-branch lines), +1 (R-branch lines), and -1 (P-branch lines) are easily derived from the symmetry properties of the Hamiltonian. Consideration of the symmetry properties of the asymmetric rotor wave functions also yields selection rules for the a-, b-, and c-type transitions allowed by the μ_a, μ_b, and μ_c components, respectively, of the dipole moment of the molecule along the principal axes. The selection rules are given in Table 1-2.

A number of computer programs such as those written by Beaudet [21], Kneubuhl and Gaumann [22], Pollnow and Chung [23], and Delfino and Ramaprasad [24] can be used to predict microwave spectra of molecular species from their known or assumed rotational constants, as well as to assign a particular spectrum to a specific molecule.

The rigid rotor approximation for calculating the rotational spectrum of a molecule has proved to be extremely useful in the interpretation of a microwave spectra. However, the nuclear framework is by no means rigid. The bond distances and bond angles of a

TABLE 1-2. SELECTION RULES FOR MICROWAVE ROTATIONAL TRANSITIONS

Dipole Moment Allowed Transition	ΔK_-	ΔK_+
μ_a	0, ±2, ...	±1, ±3,
μ_b	±1, ±3, ...	±1, ±3,
μ_c	±1, ±3, ...	0, ±2,

molecule vary as a function of its rotational state because of centrifugal distortion. This results in changes in the rotational spectrum of the molecule. The effects are usually significant for light molecules. A typical spectrum of a light molecule is usually fitted to a set of rotational constants and a number of centrifugal distortion parameters. Theoretical methods [25-27], taking into account centrifugal distortion effects, are used to calculate spectra with sufficient accuracy to reproduce the observations. This is extremely valuable in analytical applications. A detailed calculation technique, based on the Watson [25] and Kivelson-Wilson [26] model, has been presented by Kirchhoff [27]. Essentially, the Watson [25] model consists in fitting the frequencies of the observed transitions to a set of three rotational and five quartic distortion constants. The Kivelson-Wilson [26] approach would fit the microwave spectrum to nine parameters, including three rotational constants. A detailed computer program has been written by Kirchhoff [27] at the National Bureau of Standards to calculate the spectrum of a molecule taking into account centrifugal distortion and is readily available through NBS in Gaithersburg, Maryland.

The ease and accuracy with which the microwave spectrum of a molecule can be fitted to a set of effective moments of inertia or rotational constants and centrifugal distortion constants has improved the utility of MRS for analytical application. In addition, this approach has been profitably used for analysis of microwave spectra of light molecules originating from the galactic environment.

Nonrigidity in a rotor may also appear as a vibrational effect. Rotating molecules may occupy the ground vibrational state or an excited vibrational state. The spectra observed will have effective rotational constants, A_v, B_v, and C_v, appropriate to the vibrational state, v. Thus one may expect very complicated spectra if many vibrational states are excited. However, the rotational transitions belonging to the ground vibrational state, $v = 0$, are usually the most intense and hence easily assigned. The intensities of the same rotational transition in different vibrational states are related to each other by the Boltzmann distribution of population, which may be helpful in their identification. Fine structure or hyperfine structure in the rotational spectrum of a molecule may arise due to the effects of hindered internal rotation of groups such as CH_3, SiH_3, and NO_2, as well as the quadrupole electric moment of certain nuclei. These effects have implications in the use of MRS in chemical analysis but can usually be taken into account. The modified rotational spectra of molecules as a consequence of these effects can be predicted from available theoretical methods [4] in an accurate manner.

Internal rotation, that is, the rotation of one part of a molecule with respect to the remaining framework about a single bond, results in splitting of the rotational transition in certain cases. This results from the coupling of the hindered internal rotation with overall rotation of the molecule. The problem is treated in terms of a potential barrier to internal rotation of tops. For example, the rotation of the CH_3 top with respect to the remaining molecular framework CH_2-F in the CH_3CH_2F molecule results in a doublet structure. The variation of potential energy as the CH_3-top rotates with respect to the CH_2F frame can be illustrated by a potential function as shown in Fig. 1-5. The minimum and the maximum in the potential energy as the top rotates with respect to the frame corresponds to the staggered and eclipsed configurations of the molecule as shown in Fig. 1-6. There are six such equivalent positions. The appearance of the spectrum depends on the type of internal rotor and on the barrier height hindering internal rotation. Hindered internal rotation in molecules such as CH_3CH_2F, SiH_3SiH_2F, and SiH_3PH_2 internal rotor and on the barrier height hindering internal rotation. Hindered internal rotation in molecules such as CH_3CH_2F, SiH_3SiH_2F, and SiH_3PH_3 with a single threefold rotor results in splitting of certain rotational lines into doublets. The consequences of internal rotation on rotational spectra of molecules has been adequately discussed in the literature [4]. Suffice it to say, the high-resolution rotational spectrum of a molecule as affected by hindered internal rotation can usually be predicted quite accurately. In certain cases internal rotation can give rise to rotational isomers that are stable enough to give separate rotational spectra. The rotational spectra of certain molecules, such as deuterated ethyl chloride, C_2DH_4Cl, in distinctly different equilibrium configurations (see Fig. 1-7) can readily be observed and distinguished because of the high sensitivity of the moments of inertia of a molecule to its geometry.

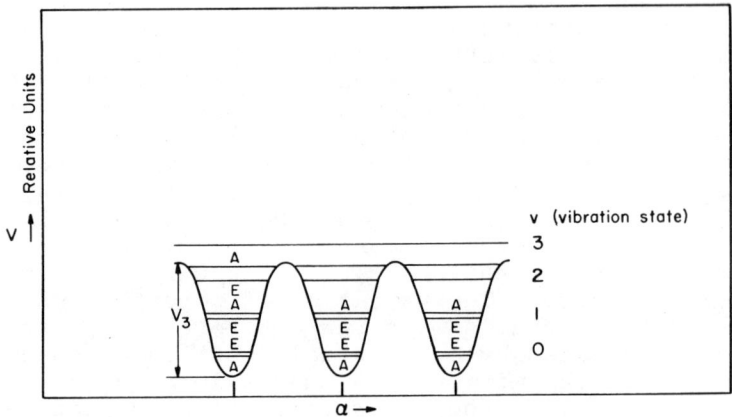

Angle of internal rotation of CH_3 top relative to the frame, CH_2F

Fig. 1-5. Torsional energy levels for rotation of CH_3 group hindered by threefold potential barrier, V. Each torsional state, ν, is split into two internal rotational sublevels, A or E.

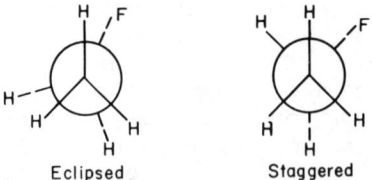

Fig. 1-6 Staggered and eclipsed configurations of CH_3 top relative to CH_2F frame in CH_3CH_2F molecule.

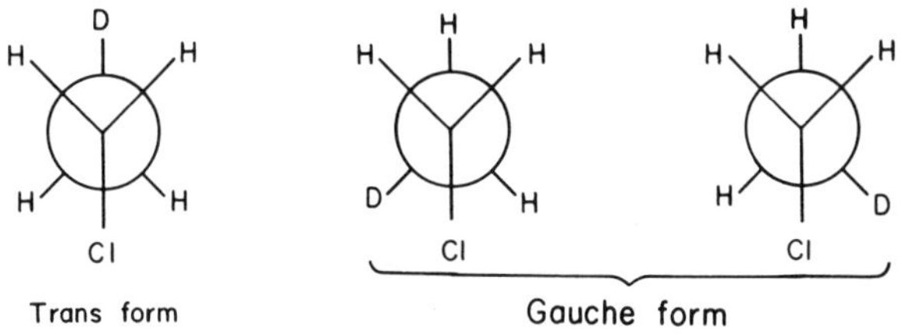

PRINCIPLES OF MICROWAVE ROTATION SPECTROSCOPY 17

Fig. 1-7. Rotational isomers of deutero-ethyl chloride.

Nuclear electric quadrupole moments arise in atomic nuclei due to nonspherical distribution of electric charge. The interaction of a quadrupolar nucleus with the external electrons leads to hyperfine structure in the rotational spectrum. In such cases the nuclear spin $\hat{I}$ couples to the resultant angular momentum, $\hat{P}$, of rotation and the resultant total angular momentum, $\hat{F}$, is quantized, with magnitude

$$|\hat{F}| = \frac{h}{2\pi} \sqrt{F(F+1)} \qquad (1\text{-}19)$$

where $F = J + I, J + I - 1, \ldots, J - I$. A particular rotational line is split into a number of components. Molecules containing one or more such quadrupolar nuclei (e.g., ^{14}N, ^{35}Cl, ^{37}Cl) have hyperfine structure in their spectra. The theory of nuclear quadrupole interaction has been worked out in detail in the literature [28]. The splitting in general decreases with J, and hence this effect has little consequence for analytical applications wherein high J lines are often measured.

Application of an electric field in the direction of the microwave electric field in the waveguide (selection rule, $\Delta M = 0$) results in a frequency shift for the rotational transitions known as the Stark effect. The frequency shift, $\Delta\nu$, is a function of the electric field applied, the electric dipole moment, and the structure of the molecule. The knowledge of intensities, spacing, and number of Stark components observed is frequently helpful in the assignment of rotational transitions of a molecule and hence has significance in terms of qualitative analytical applications. Stark effect modulation of rotational lines was first used by McAfee, Hughes, et al. [29] to increase the signal-to-noise (S:N) ratio in their microwave spectrometer. This technique is discussed in detail later on.

1-3. TECHNIQUES OF MICROWAVE ROTATIONAL SPECTROSCOPY

An analytical chemist generally considers elements such as slits, prisms, gratings, mirrors, gears, and cams as essential to the operation of a spectrometer in the infrared (IR), visible, and UV regions. Microwave spectromers are devoid of dispersive elements and optical components. In its simplest form a microwave spectrometer consists of three basic components: (1) a microwave source, (2) an absorption cell containing the gaseous sample, and (3) a detector.

Reflex klystrons, backward wave oscillators (BWO), and more recently solid-state Gunn effect diodes, are used as sources for microwave radiation. The first two devices utilize the emission of radiation by an accelerating beam of electrons. The principle of operation of a reflex klystron is illustrated in Fig. 1-8. The cathode produces a continuous beam of electrons on heating. The electrons, as they are accelerated by the electric fields toward the anode and deaccelerated by the negatively charged repeller, undergo "bunching." As a result of the oscillations in the cavity, microwave radiation is emitted. The backward wave oscillator (BWO), operating on a similar principle of bunching of electrons but without a cavity, can be electronically tuned over the entire range of a particular waveguide size of several gigahertz. Although a BWO is not as stable as the klystron as far as frequency output is concerned, use of modern methods of frequency stabilization has improved its potential to such an extent that it is currently the most extensively used commercial radiation source. Solid-state sources, on the other hand, are likely to replace both of the other types for spectroscopy up to 40 GHz because they are much less noisy, are tunable over larger bandwidths, require low-voltage supplies, and have an indefinitely long lifetime. Tunable solid-state sources, known as YIG tuned Gunn effect sources, are now available commercially and can provide continuously tunable radiation signal. Thus the S:N ratio is tremendously increased. By using phase-sensitive detection, the zero field line and its Stark components are displayed on the scope or recorder with opposite phase. The principle of Stark modulation and phase-sensitive detection is illustrated in Fig. 1-10. Observation of true line shapes requires that the Stark lobes be completely removed from the unsplit lines by application of sufficiently high peak square wave voltage. For most molecules, with the use of appropriate waveguides, a maximum voltage of 2000 is quite adequate for full modulation.

Microwave frequencies may be measured approximately by using an absorption-type cavity called the "wavemeter," which is placed between the oscillator and the detector. The wavemeter is a high-Q resonant cavity that absorbs energy from the waveguide system whenever the frequency of the incoming radiation is close to the resonance frequency of the cavity. The resonance frequency of the wavemeter can usually be varied over a wide band. With careful calibration they usually give readings correct to 1 or 2 MHz. An accurate method of measuring frequency involves mixing of the appropriate harmonic of a known reference signal, ν_m, with the microwave signal and measuring the frequency of the heterodyned radiofrequency signal leaving the mixer. The latter corresponds to the difference, $n\nu_m \pm \nu_s$, between the known reference harmonic $n\nu_m$ and the unknown microwave frequency, ν_s. A communication receiver may be used for this purpose. For example, the amplified heterodyned signals may be displayed as frequency markers on scope. By observing the value when aligned accurately with absorption lines, a precise value of the frequency of the microwave transition may be obtained. However, since most microwave oscillators are voltage tunable, it is not too difficult to phase lock them to a reference oscillator, thus "synthesizing" the desired frequency.

A typical microwave spectrometer consists of microwave sources, an absorption cell, a frequency-measurement device, a Stark modulation system, phase-sensitive detector, and a recorder. The Stark modulated microwave spectrometer originally designed by Wilson and co-workers [29] is currently being used at a number of laboratories. Commercial spectrometers such as those manufactured by Hewlett Packard (U.S.A.) and Cambridge Scientific Instruments (U.K.) are improved and automated versions of the Wilson spectrometer.

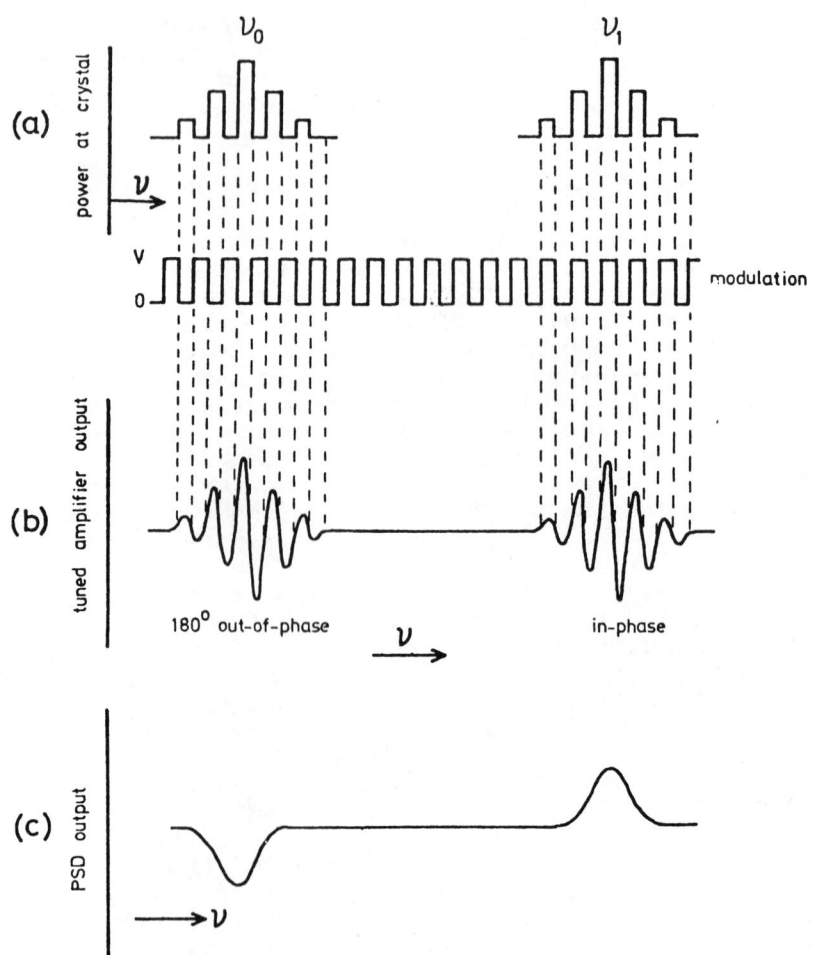

Fig. 1-10. Principles of Stark modulation and phase-sensitive detection. [With permission of Academic Press, Inc. (London) Ltd.]

1-4. INTENSITIES AND SHAPES OF ROTATIONAL TRANSITIONS

Under the conditions of thermal equilibrium the power absorbed for a transition from state 1 to state 2 from a broad spectral distribution is given by

$$\Delta P = V N_{(1)} (1 - \exp \frac{-h\nu_{12}}{kT}) B_{12} \rho(\nu_{12}) h\nu_{12} \qquad (1-21)$$

where

V	= volume of the sample, cm^3;
$N_{(1)}$	= number of molecules per unit volume in rotational state 1, cm^{-3};
k	= Boltzmann's constant;
T	= temperature, K; and
$\rho(\nu_{12})$	= radiation density, $W\ cm^{-2}$.

The Einstein coefficient of absorption, B_{12}, for the particular transition (1) → (2), is given by

$$B_{12} = (\frac{8\pi^3}{3h^2})[|<1|\mu_x|2>|^2 + |<1|\mu_y|2>|^2 + |<1|\mu_z|2>|^2] \qquad (1-22)$$

where

$$<1|\mu_x|2> = \int \psi_1^* \mu_x \psi_2 d\tau$$

and so on represent the matrix elements of the dipole-moment components of the molecule resolved along the space-fixed axes.

The power absorbed is a function of ν in the vicinity of resonance absorption by gaseous molecular species at a finite pressure in the cell. The power absorbed is maximum at $\nu = \nu_0$, where ν_0 is the resonance frequency. The line shape is not sharp but has a symmetric spread about ν_0. A given transition is characterized by its shape, Gaussian or Lorenztian, and the width, $2(\Delta\nu)$, in addition to its frequency, ν_0. The line width, $2(\Delta\nu)$, expressed in frequency units is the separation between the half-intensity points of the line (see Fig. 1-11). In contrast with the usual case in the infrared region, where the spectral slit width determines line width, the line itself limits resolution of rotational transitions

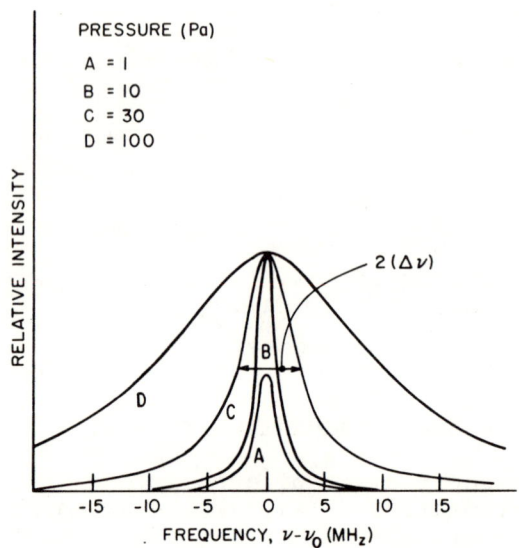

Fig. 1-11. Line shape versus frequency as function of pressure of gaseous sample.

observed in the microwave or radiofrequency range. This is apparent from the dependence of line shape on pressure of the gas molecules in the absorption cell as shown in Fig. 1-11. Note that the peak absorption at low pressures (<1 Pa) increases whereas the line width remains constant. However, at higher pressures (>1 Pa) the peak absorption remains constant but the line width increases.

A variety of factors contribute to the line width and the overall line shape. In the region of low pressures (<0.1 Pa) the line width observed is independent of sample pressures. In this pressure region the width is usually attributed to the line broadening due to the Doppler effect, collisions of molecules with the walls, and modulation broadening. The width of a Doppler broadened line can be expressed by the relation

$$2(\Delta \nu) = 7.15 \times 10^{-7} \; \frac{T}{M}^{1/2} \; \nu_0, \qquad (1-23)$$

INTENSITIES AND SHAPES OF ROTATIONAL TRANSITIONS

where M is the molecular weight in atomic mass units and T is the absolute temperature. For example, in the case of the OCS absorption at 24 GHz, near room temperature (T = 300 K) the Doppler width, $2(\Delta\nu) \sim$ 38 kHz. Doppler broadening gives a Gaussian line shape.

At low pressures (<0.1 Pa) the collisions of molecules with walls become more important than the intermolecular collisions. For a rectangular waveguide with the inside cross-section dimensions of a and b and length c in cm, the line-width broadening due to wall collisions can be approximated by the relation [30]

$$2(\Delta\nu) = 10[\frac{1}{a} + \frac{1}{b} + \frac{1}{c}] \frac{T}{M}^{1/2} \text{ kHz.} \qquad (1-24)$$

One calculates the line width due to wall collisions in the case of the 24-GHz OCS transition, when T = 300 K in a long K-band waveguide (a = 0.432 cm, b = 1.064 cm) as ∿70 kHz.

Broadening due to the fact that the time required for the adjustment of molecular wave function in the presence of Stark modulation, is of the same order of magnitude as the molecular collision relaxation time, τ, and can be estimated from the theory discussed by Karplus [31]. The line width of an OCS line due to the modulation effect is about 1.2 times the modulation frequency. Since 100 kHz modulation frequency is often used, this is a significant source of line broadening. This kind of broadening can be reduced by operating the spectrometer at low (e.g., 5 kHz) Stark modulation frequency. However, the detector crystal noise of point-contact diodes varies inversely with the modulation frequency, and thus the sensitivity is better at higher modulating frequency unless special detectors such as back-diodes are used.

The broadening due to the wall-collision and modulation effects both give a Lorentzian line shape. However, after Doppler effects are included the low-pressure line shape is close to Gaussian.

Molecular collisions cause line broadening as the pressure increases (>0.1 Pa). The width in Hz is given by

$$2(\Delta\nu) = \frac{1}{\pi\tau}, \qquad (1-25)$$

where τ is the mean time between molecular collisions that tend to restore equilibrium in the cell.

Normally the operating pressure of an analytical spectrometer is such that the dominating line shape is given by the Lorentzian function [32]

$$S(\nu,\nu_0) = \frac{\Delta\nu}{\pi[\Delta(\nu-\nu_0)^2 + (\Delta\nu)^2]} \tag{1-26}$$

The collision time, τ, is inversely proportional to pressure over pressure range of 1 to 1000 Pa, during which interval $2(\Delta\nu)$ increases linearly with pressure. In fact, whereas peak intensity is independent of pressure, the half-width $(\Delta\nu)$ varies directly as pressure,

$$\Delta\nu = Kp, \tag{1-27}$$

where K is a temperature-dependent parameter.

When the incident microwave radiation density is sufficient to induce molecular transitions from the lower state (1) to the upper state (2) more rapidly than thermal equilibrium can be restored by various relaxation mechanisms, the peak intensity decreases and the line width increases. This effect, called "power saturation," was illustrated in Fig. 1-2, Case II. The saturation effect has been discussed by Townes [3] in some detail. The situation arising from power saturation in a microwave spectrometer, such as is observed when samples at low pressures are exposed to increasing radiation power, is illustrated by Fig. 1-12. This figure shows that the absorption coefficient, α_ν [defined in Eq. (1-28)], relative to its coefficient at the resonance peak, α_{ν_0}, decreases with increasing power. However, the signal (for a linear detector) is nearly proportional to α_{ν_0} at low power levels, but not at higher powers. This indicates that reliable measurements of intensity can only be made at low-power level in the cell unless power-saturation effects are taken into account.

Unresolved nuclear quadrupole and hindered internal rotation hyperfine structure, overlapping lines, and incomplete Stark modulation are very often causes of distorted line shapes. Certain deconvoluting procedures can be used for lines with unresolved components arising from the nuclear quadrupole effect and hindered internal rotations.

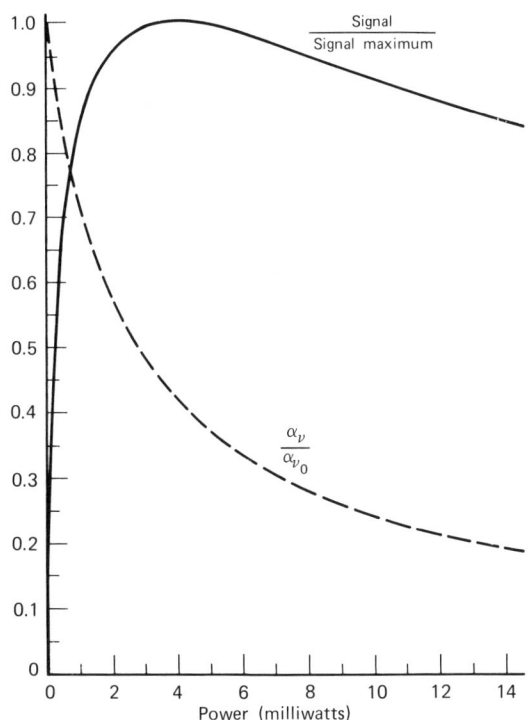

Fig. 1-12. Measured values of spectrometer signal amplitude and intensity coefficient as function of radiation power level. A transition in sulfur dioxide SO_2 at 9403.22 MHz was used. The sample pressure was 0.1 Pa at room temperature. (With permission of the American Institute of Physics.)

1-5. ABSORPTION COEFFICIENT, α_ν, AND THE INTEGRATED ABSORPTION, α_{int}

Based on the formula derived by Van Vleck and Weisskopf [33] and Coles [34], the absorption coefficient, α_ν, expressed in cm^{-1} for a rotational transition $J, K_-, K_+ \to J', K_-', K_+'$ observed for an asymmetric molecular species in the microwave absorption cell at pressures of 0.1 to 100 Pa, is given by the equation

$$\alpha_\nu = \frac{8\pi^2 N F_{J,K_-,K_+} \nu^2 \mu_F^2 \lambda_F}{3ckT(2J+1)} \cdot \left[\frac{\Delta\nu}{(\nu-\nu_0)^2 + \Delta\nu^2}\right],$$

(1-28)

where the second term on the right-hand side of the equation represents the Lorentzian line-shape function in the low-pressure region. Furthermore,

- N = number of molecules per cm^3;
- ν = frequency of microwave radiation;
- ν_0 = resonance frequency at which the absorption is maximum;
- $\Delta\nu$ = half-width of the line at half-intensity points;
- c = speed of light;
- μ_F = electric dipole moment component responsible for the transition (where F = a, b, or c);
- k = Boltzmann constant; and
- T = temperatures in degrees Kelvin.

The line strengh, λ_F, for the asymmetric top transition, $J,K_-, K_+ \rightarrow J', K_-', K_+'$, is related to the dipole-moment matrix element [defined in Eq. (1-22)] by the relation

$$\lambda_F = |<J,K_-,K_+|\mu_F|J',K_-',K_+'>|^2 \frac{2J+1}{\mu_F^2}$$

(1-29)

Line strengths and frequencies of transitions for a large number of molecules have been tabulated in the NBS Microwave Spectral Tables [35]. As information becomes available for other molecules, this compilation will likely be updated. Additional frequency catalog data on a number of interesting molecules for analytical MRS applications are available in technical reports [36] and published literature [37]. Such tables can be of considerable help in applications of MRS to chemical analysis.

The fraction, F_{J,K_-,K_+}, of the total molecular population occupying the lower rotational state, J,K_-,K_+, in the ground vibration state V, (V = 0) is given by the equation

$$F_{J,K_-,K_+} = (2J+1) F_v g \cdot \frac{\exp^{-(\varepsilon_1 - \varepsilon^{(0)})/kT}}{Q_r},$$

(1-30)

where

ε_1 = the energy of the lower rotation state, J,K_-,K_+, of the transition;

$\varepsilon_1^{(0)}$ = the energy of the lowest state available to the molecule;

g = the statistical weight factor of the level (the factor is unity except when the molecule contains identical nuclei occupying identical symmetrical positions in the molecule);

Q_r = the rotational partition function; and

F_v = the fraction of molecules in the vibrational state v, (v = 0).

The approximate dependence of rotational partition function, Q_r, on rotational constants for linear, symmetric, and asymmetric tops, respectively, can be expressed by the following proportionalities:

$$Q_r : \frac{T}{B} \quad \text{(linear)}$$

(1-31)

$$Q_r : (\frac{T^3}{AB^2})^{1/2} \quad \text{(symmetric)}$$

$$Q_r : (\frac{T^3}{ABC})^{1/2} \quad \text{(asymmetric)},$$

where T is the absolute temperature. The fraction, F_v, is related to the vibrational energy and the vibrational partition function of the molecule and is available in most books on spectroscopy.

The expression for the peak absorption coefficient, α_{ν_0}, can be derived readily from Eq. (1-28) by letting $\nu = \nu_0$,

$$\alpha_{\nu_0} = \frac{8\pi^2 N F_{J,K_-,K_+} \nu_0^2 \mu_F^2 \lambda_F}{3ckT(2J+1) \cdot \Delta\nu} \quad cm^{-1}.$$

(1-32)

By putting $N = (xp/kT)$ (x = mole fraction of the absorbing molecular species and p = total pressure of gaseous species in the cell) can be expressed as

$$\alpha_{\nu_0} = \frac{U \cdot \nu_0^2 \cdot x \cdot p}{\Delta \nu} \text{ cm}^{-1}, \qquad (1\text{-}33)$$

where

$$U = \frac{8\pi^2 F_{J,K_-,K_+} \mu_F^2 \lambda_F}{3ck^2T^2(2J+1)} \text{ s}^3 \text{ gm}^{-1}.$$

At a given temperature, U is constant for a particular transition of the particular molecular species. Then, since $\Delta \nu : p$ [Eq. (1-27)] the pressure cancels in Eq. 1-33), and the peak absorption coefficient in the pressure range ~ 1 Pa to 1000 Pa is independent of total pressure.

Another useful parameter, particularly for quantitative analysis, is the integrated absorption coefficient, α_{int}, given by the equation

$$\alpha_{int} = \int_0^\infty \alpha_\nu \cdot d\nu = \pi \cdot \Delta \nu \cdot \alpha_{\nu_0} \text{ cm}^{-1}\text{s}^{-1}. \qquad (1\text{-}34)$$

The expression for ΔP given in Eq. (1-21) can be related to α_{int},

$$\alpha_{int}, \Delta P = \alpha_{int} \ell \left(\frac{\lambda}{\lambda_g}\right) P_a, \qquad (1\text{-}35)$$

where ℓ is the cell length, λ is the free-space wavelength, λ_g is the wavelength in the waveguide cell, and P_a is the time averaged power.

It appears from Eq. (1-34) that measurement of the area under the absorption line will yield α_{int}. Examination of Eq. (1-33) leads to the conclusion that α_{int} is independent of line width and is linearly dependent on the partial pressure of the component in a given mixture. Thus, this method, though difficult to apply, can give direct absolute quantitative results.

In any critical examination of the potential of microwave rotational spectroscopy for analytical applications, one must consider whether the absorption spectrum of a particular molecule is detectable and if its spectral intensity can be measured. The stronger the absorption by molecules, the more readily the chemical species is detectable in small concentrations in a mixture of gases and the wider the range of concentrations over which it can be determined quantitatively.

A number of factors from the foregoing discussions should be considered in connection with applications in analytical determinations. The half-width, $\Delta\nu$, is constant in the low-pressure (<0.1 Pa) region. Increasing the concentration of the absorbing molecules in the cell will increase the peak absorption coefficient. At higher pressures (1 to 1000 Pa), $\Delta\nu$ increases with increase in pressure, but the coefficient α_{ν_0} stays constant (see Fig. 1-11).

Equations (1-33) and (1-34) suggest several ways to obtain quantitative information from rotational absorption measurements. An absolute measure of N, the number of absorbers per unit volume, can be made by measuring the area under the absorption signal of a completely spectrally isolated transition, that is, one having a well-defined zero absorption baseline. Another method would be to measure both the power absorbed at the resonance peak and the half-width of the line at half maximum height. The product of these two values is proportional to N. It is sometimes experimentally possible to make the half-width a constant of the measurement. Therefore, a third method would be to measure only the maximum power absorbed at the resonance peak while keeping the line width constant. Still another method would be to compare the peak height of an unknown amount of absorber with a known amount of either the same or a different absorber with similar absorption coefficient.

In practice, all of these measurements are somewhat difficult to make. Each of the methods are discussed in detail in Chapter 5. It should also be noted that the absorption coefficient of molecules are functions of temperature in a complicated manner, as is readily apparent from an examination of Eqs. (1-28), (1-30), and (1-31). Thus temperature must be controlled for accurate quantitative determinations.

Because the square of dipole-moment component appears in the numerator of the expression for the absorption coefficient of a molecule [see Eq. (1-28)], there is a definite advantage in detection and intensity measurements of rotational transitions of molecules with large electric dipole moments. Many hydrocarbons with their very small dipole moments can be detected only with difficulty. The partition function, Q_r, is a function of moments of inertia of a molecule, being larger for large molecules. Therefore, large molecules will generally have weak absorptions. Furthermore, due to distribution of population into different vibrational states v (v_1, v_2, ..., etc.) in the case of large molecules containing structural units with low-frequency vibrational states, the spectra usually are weak. This makes chemical analysis of extremely large molecules quite difficult. Because of v^2 dependence of α_v, the stronger lines in general are observed at higher frequencies. This is, in fact, very valid for practical analytical determinations. The typical spectrometer frequency range for analytical applications is the R-band region (26.5 to 40 GHz).

As was mentioned earlier, power saturation effects in the microwave region are important, and one should be aware of those effects in analytical determinations. Under conditions of power saturation it is difficult to relate the usual experimentally observed α_{v_0} or α_{int} to the concentration of absorbing molecules. With excessive radiation density, $\rho(v_{12})$, in the cell [see (Eq. 1-21)], the thermal relaxation process returning the excited molecules from states (2) to (1) cannot compete successfully with the excitation of molecules in states (1) to (2). Hence the population difference in levels (1) and (2) is no longer given by the Boltzmann distribution law (see Fig. 1-2, Case II). This results in a decrease in the peak absorption coefficient, α_{v_0}, as defined for the usual low radiation-density level in the cell. The effective peak absorption coefficient, α_v, can now be expressed by the equation [38],

$$\alpha_v = \frac{\alpha_{v_0}}{1 + KP_0} \text{ cm}^{-1}, \qquad (1\text{-}36)$$

where K is the power-saturation coefficient and P_0 is the incident power expressed in milliwatts. Harrington [38,39] gives the power-saturation coefficient K by the equation

$$K = \frac{\lambda g}{\lambda} \frac{64\pi^3 |\mu_{12}|^2 \tau \cdot t}{3h^2 cA} \text{ mW}^{-1}, \qquad (1\text{-}37)$$

where

λ_g/λ = ratio of waveguide to free-space wavelengths;

μ_{12} = dipole-moment matrix elements for the transition (1) to (2), in debye units;

τ = mean time in seconds between collisions that broaden the line;

t = mean time in seconds between collisions that restore thermal equilibrium; and

A = area of the cross-section of the cell, cm^2.

The fractional loss of radiation under the conditions of power saturation at the transition frequency may be considered to be proportional to the product of concentration, N, of the absorbing species and the broadening relaxation time, τ. It is difficult to interpret the microwave-intensity data under power saturation because of inseparability of N from τ. Harrington [38,39] has developed the theory and techniques to solve this problem. The microwave signal received in the center of the waveguide at the detector of a spectrometer is proportional to the electric field change, ΔE. Since power is proportional to electric field squared, E^2, Eq. (1-35) can be written as

$$\Delta E_g : \alpha_\nu \ell P_0^{1/2}. \qquad (1-38)$$

Harrington defines a new intensity coefficient, Γ, applicable in the case of power saturation by the equation,

$$\Gamma = \alpha_\nu P_0^{1/2} = \frac{\Delta P}{\ell P_0^{1/2}} = \eta\phi. \qquad (1-39)$$

Examination of Eqs. (1-38) and (1-39) reveal that Γ is proportional to the microwave signal due to absorption by the gaseous molecules and may be determined by use of suitable techniques to be described later.

The quantity η is given by the equation

$$\eta = \frac{\alpha_{\nu_0}}{K^{1/2}} W^{1/2} \ cm^{-1} \qquad (1-40)$$

and depends directly on the number of absorbing molecules, N, present, but unlike α_{ν_0} it is not affected by the presence or absence of other molecular species. The quantity ϕ, is a dimensionless function of the product KP_0 and is given by the equation

$$\phi = (KP_0)^{-1/2}[1-(1 + KP_0)^{-1/2}], \qquad (1\text{-}41)$$

or for a Stark modulated rectangular cell operating in the TE_{10} mode by a more complicated expression given in a later experimental section devoted to analytical determination.

The power-saturation coefficient K depends on the sample composition and on τ and t and thereby on total pressure. The function ϕ is itself a function of (KP_0), and hence ϕ can be maximized by adjusting the value of K and P_0. However, η is independent of P_0. Therefore, it is always possible to reach the maximum value of ϕ on a ϕ versus (KP_0) plot by increasing the value of P_0. It can be shown that this value of P_0, namely, P_s, in fact corresponds to maximum signal amplitude, S. Furthermore, according to Eqs. (1-33), (1-37) and (1-40), η can be written as

$$\eta : (\tfrac{\tau}{t})^{1/2} xp , \qquad (1\text{-}42)$$

where x is the mole fraction of the particular absorbing molecular species and p is the total pressure in the cell, respectively. A reasonable assumption can be made that τ is equal to t so as to reach an equilibrium and hence,

$$\text{signal}: \Gamma_{max}: \eta\phi_{max}: xp . \qquad (1\text{-}43)$$

CHAPTER

2

INSTRUMENTATION

2-1. COMMERCIAL SPECTROMETERS

2-1-a. Hewlett Packard Model 8460A, Microwave Rotational Resonance Spectrometer

Hewlett Packard Company developed a research microwave instrument through the late 1960s and put on the market a fully automated microwave spectrometer [40,41], the Hewlett Packard (HP) Model 8460A, in the year 1970. The block diagram of this spectrometer is shown in Fig. 2-1. The spectrometer (no longer made) consisted of interchangeable sources of microwave radiation, a sample absorption cell and a bridge-type waveguide system, the calibration arm, the detector system, and a strip-chart recorder.

This instrument was particularly designed to fill the needs of the microwave research community; its analytical utility was of secondary emphasis. An applications group at Hewlett Packard Company was making an effort to demonstrate its more general analytical utility about the time of the company's decision to drop the instrumentation from its product line in 1974.

Backward wave oscillators in the frequency range 8.0 to 40.0 GHz are used as sources of monochromatic radiation in the HP Model 8460A. The BWO source is stabilized by phase locking its output to a harmonic of a 400 to 420-MHz reference oscillator. The reference oscillator output frequency is programmable in steps of 1 Hz so that the frequency of the BWO can be varied in steps of 100 Hz. Since the smallest scan rate is rapid compared to the frequency steps, the scan approximates a continuous sweep. Separate BWOs are used as radiation sources in the four frequency bands in the 8.0 to 40.0 GHz range. The BWOs for the

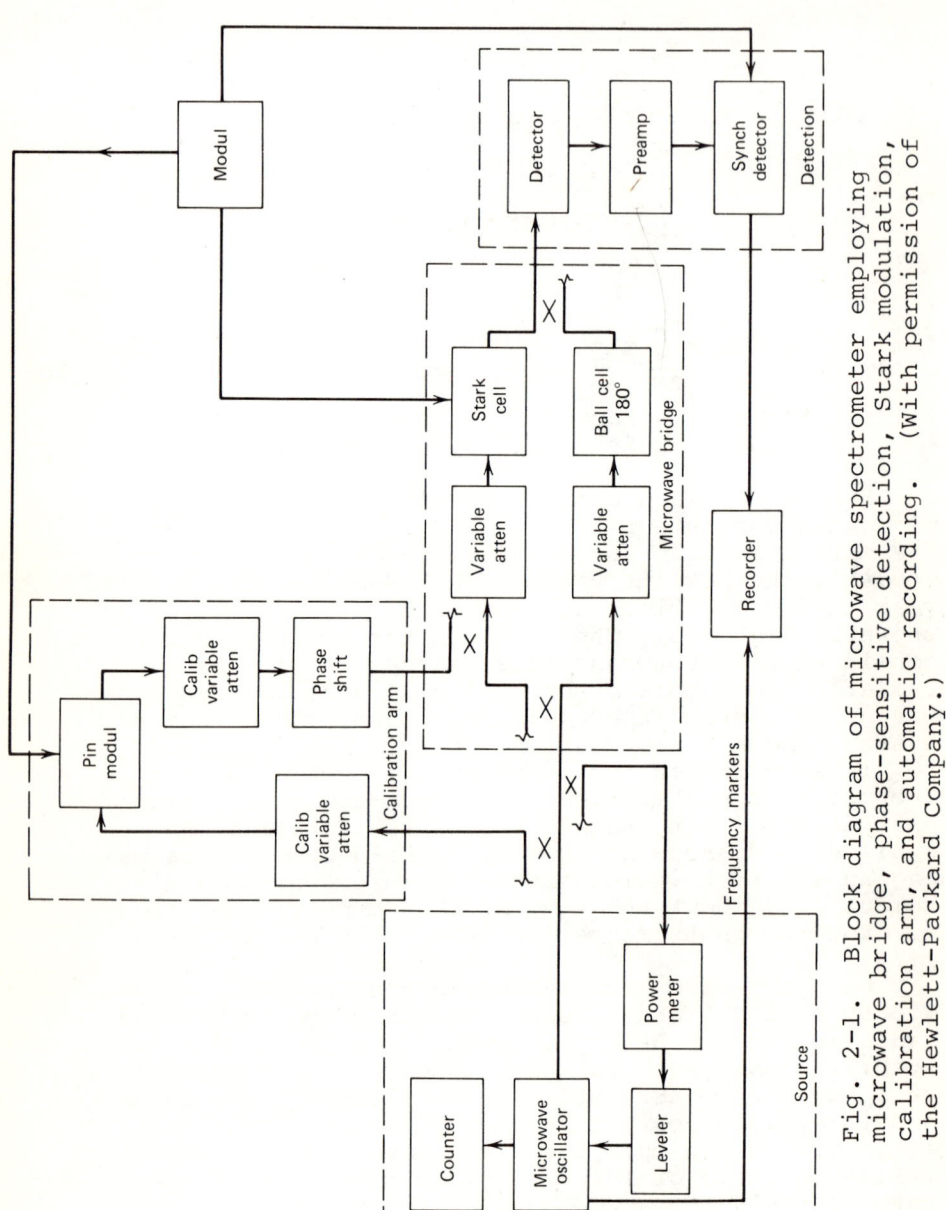

Fig. 2-1. Block diagram of microwave spectrometer employing microwave bridge, phase-sensitive detection, Stark modulation, calibration arm, and automatic recording. (With permission of the Hewlett-Packard Company.)

X-band (8 to 12.4 GHz), P-band (12.4 to 18 GHz), K-band (18 to 26.5 GHz), and R-band (26.5 to 40 GHz) are supplied as plug-in units and can be installed easily. The source may be swept over part or all of the frequency range of any band. The limits of frequency range as well as rate of scan can be selected by dial switches on the front panel of the controller. For example, a spectrum can be scanned at any one of the 13 different scan rates, ranging from 0.001 MHz to 10 MHz per second. The frequency is continuously displayed with 0.001-MHz resolution on a digital counter.

The absorption cell, of 0.5ℓ approximate volume, consists of two X-band Stark cells gold plated on the inside, each 1 meter in length. The rough vacuum for the cell is provided by a mechanical pump, with higher vacuum achievable with an ion pump. A zero-based square wave electrical field, alternating in the 0 to 2000 V/cm range at 33.333 kHz frequency, is applied to the molecules by the Stark septum located in the middle of the Stark cells. A cross-sectional view of the Stark cell and the septum is shown in Fig. 2-2. The absorption at the transition frequency of a polar molecule is modulated at the Stark modulation rate. The resulting modulated power is received by a point-contact wide-band silicon diode. The output is fed to a tuned preamplifier and then to a phase-sensitive detector that is synchronized with the modulation frequency. The spectrum can be displayed on a strip-chart recorder. The frequencies are marked on the chart automatically at chosen intervals such as 0.01, 0.1, 1.0, 10.0, or 100.0 MHz.

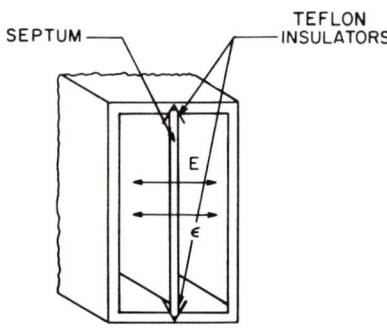

Fig. 2-2. Cross-section of Stark cell and septum.

A typical display of the zero-field line and the Stark component (appearing in opposite phase) is presented in Fig. 2-3 for the J = 0→1 transition of isotopic OCS in natural abundance. The microwave power entering the absorption cell essentially all reaches the detector except for slight attenuation in power level due to molecular absorption for the "in line" Stark cell operation of the HP spectrometer.

Certain unique features available with the HP system have been extremely valuable for application work in the field of chemical analysis. The bridge-type waveguide system enables one to vary the power level in the Stark cell considerably and yet maintain a constant current at the crystal detector, thus providing a means for power-saturation measurements. Such an arrangement [41] of the spectrometer is illustrated in Fig. 2-4. The microwave power is divided into two portions that travel (180° out of phase) through the "absorption cell" and the "balance cell," respectively. The microwave radiation emerging from the two arms are combined before being received by the detector. The net power is the difference in the

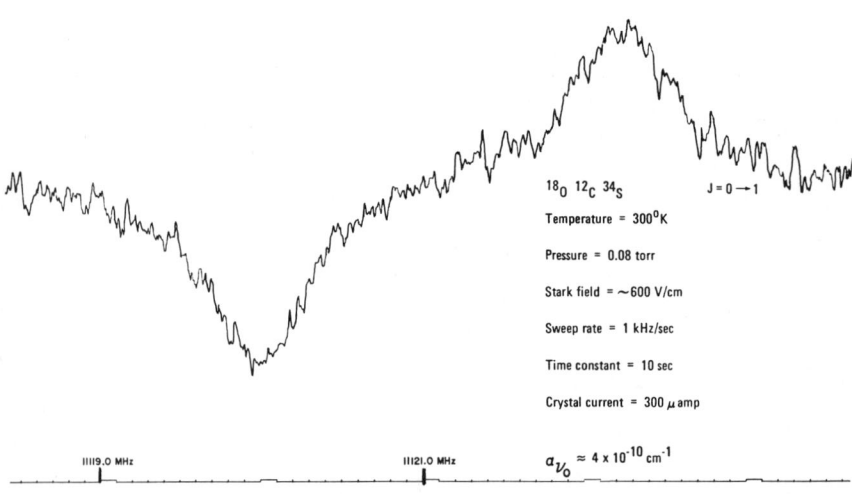

Fig. 2-3. Zero field line and Stark component for J = 0→1 transition of OCS. (With permission of the Hewlett-Packard Company.)

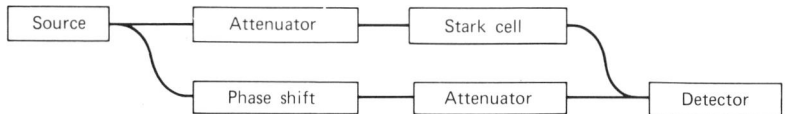

Fig. 2-4. Bridge arrangement in HP Model 8460A Stark modulated microwave spectrometer.

powers from the two arms and can be adjusted to any arbitrary value with the attenuators and phase shifter. This provides a means to balance out any amount of excitation power in the Stark cell so that the detector sees only enough power to optimally bias it. The calibration arm, consisting of a phase shifter, a calibrated attenuator, and a PIN-diode, is another convenience proven valuable in analytical determinations. The PIN-diode acts as a very fast switch to attenuate microwave power. It can thus be used to simulate an absorption for calibration of actual absorptions. These new features have been effectively utilized for analytical determination of chemical species, as is discussed later.

One of the most advantageous features with the HP system has been the ease with which the spectrometer can be fully interfaced with the computers. White [42], who achieved higher resolution and sensitivity in his measurements by the HP Model 8460A spectrometer interfaced with a HP Model 2116B digital computer, has described examples of signal averaging, digital filtering, and data reduction. Scharpen and Rasukolb [43] used a similar system for quantitative analysis of a gaseous mixture.

2-1-b. Cambridge Scientific Instruments
        Analytical Microwave Spectrometer

In the early 1970s, a British company, Cambridge Scientific Instruments Limited (CSI), developed a microwave rotational spectrometer specifically for analytical applications. In this apparatus the designers tried to solve some major problems regarding the use of the technique for quantitative measurements. Although the resulting instrument was generally analytically oriented, the company has had poor success in marketing it, quite likely because of a combination of high cost and sophistication. The instrument is no longer made.

The block diagram of this instrument [44] is shown in Fig. 2-5. It consists of three major components, a sample handling system, the basic spectrometer, and the data-processing system. The CSI spectrometer has incorporated only the R-band (26.5 to 40.0 GHz) in its design. The BWO can be set to scan between any two frequencies in the range 26.5 to 40 GHz. Convenient slow, medium, and fast scans may be used to record any spectrum. The square wave Stark voltage alternates at 40 kHz between 0 and 2 kV. The Stark modulated signal is amplified by a tuned amplifier and detected by a phase-sensitive detector.

Precise measurement of integrated intensities, a useful determination for analytical applications, depends on reliable measurement of area under an absorption line. As is explained in a later section, requirements such as good separation of Stark lobes from the zero field line, avoiding interference from joining lines or Stark lobes belonging to neighboring lines, and an acceptable S:N ratio for the observed transition, have to be met in any measurement of area under the absorption spectral line. The CSI design uses an automatic electronic integration system that measures the line area between the peak point and a point below which is a known fraction of peak height above the baseline. Assuming Lorentzian shape for the line, the area between these points is arranged to give directly the true area under the curve. This integrator device gives the instrument a feature that is extremely useful in analytical determinations. The CSI spectrometer has been designed to allow simple interfacing to a computer for logging spectra automatically. Output to the computer is derived from the integrator device and from frequency markers that can enable determination of peak areas at different regions of the frequency scan. The full potential of this facility are explored in a later section.

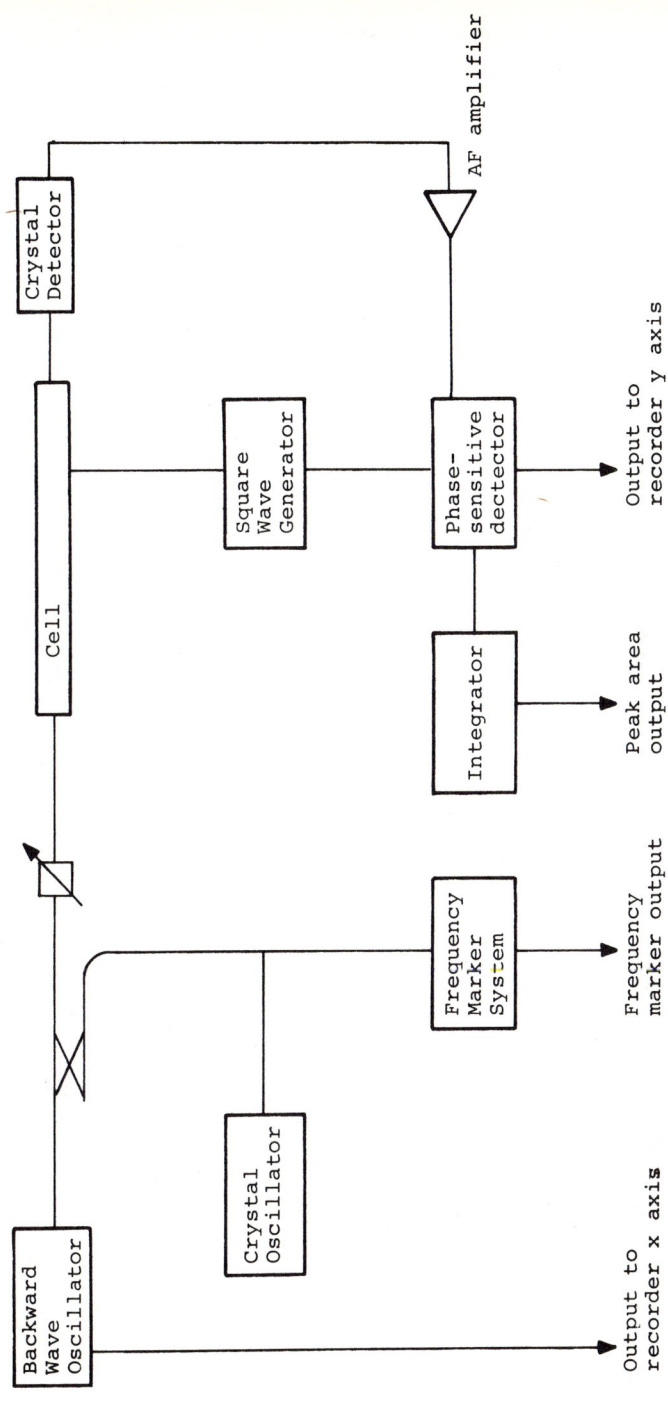

Fig. 2-5. System diagram of Stark Modulated microwave spectrometer designed by CSI. (With permission of Cambridge Scientific Instruments Ltd., Cambridge, U.K.)

2.2. MILLIMETER WAVE SPECTROMETERS

Microwaves in the wavelength regions from 10 mm to 1 mm and from 1 mm to 0.3 mm are usually called "millimeter waves" and "submillimeter waves," respectively. Millimeter and submillimeter wave spectrometers yield similar information about molecular species in the vapor phase, as discussed in an earlier section of this text.

Detection, characterization, and quantitative analysis of a light molecular species, that is, those having high rotational energies, may be conveniently conducted [45] using a spectrometer operating in the wavelength range 0.3 to 10 mm.

2-2-a. Video-type Millimeter Wave Spectrometer

The block diagram of a modern video-type spectrometer [45,46] operating in the millimeter wave region is shown in Fig. 2-6. Harmonics of the fundamental frequency from klystrons and BWOs generated by a point-contact rectifying diode produce the coherent radiation in the millimeter and submillimeter wavelength regions. The millimeter wave radiation is channeled by horns and dielectric lenses into an oversized free-space absorption cell. The power level in the cell is monitored by the use of a point-contact diode for the millimeter waves, similar in design to one used for harmonic generation. A photoconductive indium antimonide detector operated at 1.6 K is used for detection. Sensitivity of the indium antimonide detector operated at 1.6 K is at least one order of magnitude higher than the usual point-contact diode detector for the spectral region.

However, other techniques such as the Stark effect-modulation, source frequency modulation, or saturation-effect modulation [47], can also be used in millimeter wave spectrometers to enhance the S:N ratio. The free space cells such as those used in millimeter wave spectroscopy (see Fig. 2-13) are quite suitable for the study of reactive as well as high-temperature molecules. Considerable enhancement in sensitivity of the spectrometer has been achieved through the use of a digital computer for signal averaging, digital filtering, and data reduction. This has been shown for the millimeter wave region by Winnewisser [48]. The S:N ratio is improved in a

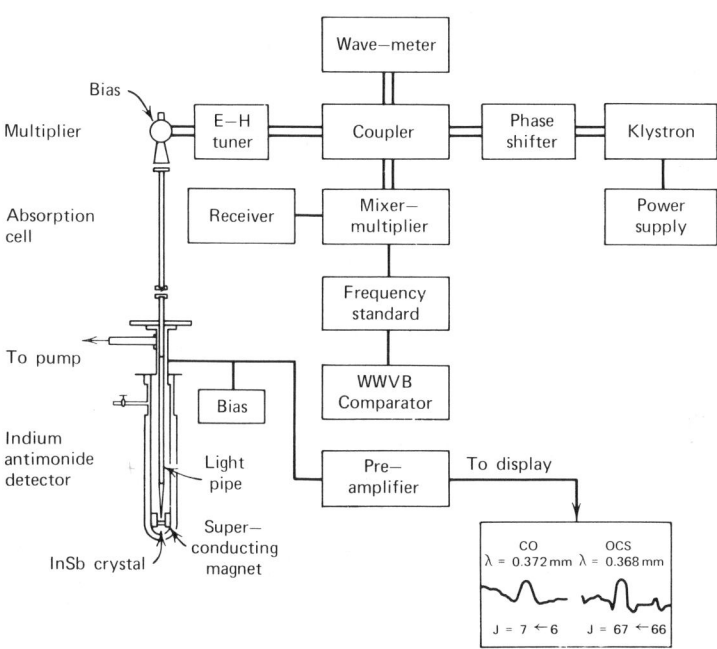

Fig. 2-6. Block diagram of video-type millimeter wave spectrometer. (Reproduced by courtesy of The American Institute of Physics and Academic Press.)

video-type spectrometer by sweeping the frequency over the absorption many times in a second and detecting the absorption on an oscilloscope. The signal-averaging technique in its application allows an improvement of the S:N ratio by a factor of 10^2 using 10,000 sweepts. A limiting sensitivity for detecting α_{ν_0} equal to 10^{-8} cm^{-1} is obtainable by this technique.

2-2-b. Millimeter Wave Fabry-Perot Spectrometers

Semiconfocal Fabry-Perot cavity spectrometers [49,50] are used to achieved high S:N ratios in the millimeter wave region, where less power is available. In comparison to the parallel-plate cavities, these are less critical to align and diffraction losses are reduced considerably. The open structure of the cavity is suitable for conducting measurements on reactive molecules such as free radicals (see Section 2-4 for the description of such a spectrometer).

2-3. COMPUTER INTERFACING OF MICROWAVE SPECTROMETERS

Numerous researchers have successfully interfaced spectrometers operating in the centimeter and the millimeter wavelength regions with digital computers with significant gains. The marketing of commercial microwave spectrometers suitable for computer interfacing, such as the HP Model 8460A and the CSI model, as well as the availability in general of a variety of electronic components, such as the BWOs, phase-locked oscillators, frequency synthesizers, and analog-to-digital converters, have indeed made such computer interfacing popular. Such systems have been developed by Gwinn, Luntz, et al. [51] and White [42]. Recently, computer-controlled microwave spectrometers have also been reported by Woods and Dixon [52] and Winnewisser [48] in the centimeter and millimeter wavelength regions. Haase, Gegenheimer, et al. [53] have developed a computer-interfaced microwave spectrometer in the centimeter wavelength region for detection of air contaminants in trace amounts.

Chemical analysis in a routine manner on a computer-interfaced microwave spectrometer is feasible because of availability of features such as the collection, reduction, and analysis of spectral data in an automated manner. An example of this is described in Chapter 5.

Weak rotational transitions with absorption coefficients as small as 10^{-9} cm^{-1} or less are beyond the limits of sensitivity using conventional techniques. However, interfacing of a microwave spectrometer with a computer allowed Haase, Gegenheimer, et al. [53] to measure transitions of extremely small absorption coefficients (i.e., $<10^{-9}$ cm^{-1}; see Fig. 2-7).

In addition to controlling the operations of the spectrometer, the computer performs functions such as digital averaging of a large number of successive scans, and smoothing through digital filtering. Interfacing of a microwave spectrometer with a

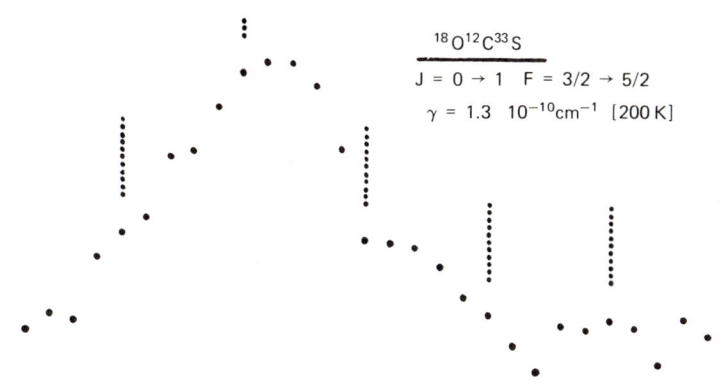

Fig. 2-7. Spectrum trace for transition; $J = 0 \rightarrow 1$, $F = 3/2 \rightarrow 5/2$, of the $^{18}O^{12}C^3S$ molecule. (With permission of authors, W. Zeile, B. Haase, et al. [53]).

computer has considerably increased the usefulness of MRS for chemical analysis. Gwinn, Luntz et al. [51] have discussed such an interface in detail. They used this system to control the spectrometer operations, and to acquire and reduce data.

The advantages of a computer-controlled spectrometer can be best illustrated through an example. The computer-controlled spectrometer described by White [42] is discussed to illustrate the principle of a typical interface. The HP Model 8460A was interfaced with a HP Model 2116B digital computer, which has a disk memory. The 16 input/output channels operate on a priority interrupt basis for interface with the disk memory, teletype, and paper-tape equipment and for control of spectrometer functions. Voltage-programmed signals can be supplied by the computer to the spectrometer, and the computer can receive voltages from the spectrometer. This facility is provided by the digital-to-analog converter and an analog-to-digital converter. The interface uses a programmable microwave source. The microwave power level is controlled over a 10-dB range in 1-dB steps. Furthermore, the microwave signal is tunable over the waveguide band in 0.001-MHz steps. The power-leveling amplifier compares the microwave power from the spectrometer with its own reference voltage and generates an appropriate voltage that then controls the BWO power output. The control of power level over the 10-dB range is achieved by programming the reference voltage of the amplifier by a single-digit BCD output. A backward wave oscillator can be tuned to any preset frequency simply by adjusting the helix voltage of the tube. The helix tuning characteristics of the particular tube in use are stored in the disk. The proper helix voltage is chosen by the computer and applied to the BWO to tune the oscillator to the proper frequency. The frequency is controlled in three steps. The first BCD output tunes the BWO frequency with 1000-step resolution over the range 26.5 to 40 GHz to a range within the capture potential of the phase synchronizer. The frequency of the reference oscillator, tunable in the range 400.00 to 420.00 MHz in 1-Hz steps, is adjusted by a suitable BCD output to generate a harmonic approximately 20 MHz below the desired microwave frequency. The harmonics are then mixed with the microwave source frequency. The resulting beat frequency is compared by the synchronizer with the intermediate-frequency reference frequency and the subsequently generated correction voltage is applied to the BWO to adjust the microwave source frequency in a manner such that the beat and intermediate-frequencies are the same. The third BCD output fine tunes the frequency of the microwave signal. The Stark voltage may be controlled manually or by the computer. The Stark ground-to-base (0 to 2000 V) and base-to-peak (0 to 2000 V) voltages are controlled in 8-V steps in the Stark voltage range through a 0 to 10-V d.c. output from the computer. The signal amplifier gain may also be controlled by the computer. Data measurements are made by the voltage-to-frequency converters. The spectrometer operating status, such as Stark voltage, power level at the detector, and sample pressure are continuously monitored by the computer.

The S:N ratio for microwave absorptions measured on a conventional spectrometer are improved by employing Stark effect modulation and phase-sensitive detection in this spectrometer. An electronic filter (i.e., resistance-capacitance noise filter on the amplifier output) is helpful in the improvement of S:N ratio but may result in distorted line shapes if the source is swept too rapidly.

Significant improvement in the S:N ratio for microwave absorptions are achieved by use of the new techniques with a computer-controlled microwave spectrometer. The time integration of signals using a voltage-to-frequency converter results in the reduction of noise and in improved resolution of the signal. The S:N ratio can be further improved through digital filtering and digital averaging. Computer filtering involves mathematical curve smoothing. A least-squares fit of a general polynomial function of the form $y = a_0 + a_1x + a_2x^2 + a_3x^3 + \cdots$ to a limited number of odd data points is arranged. The polynomial function is chosen because it is the best smooth curve that can represent the data in the small spectral region. The value of the intensity at the central data point is calculated by using the polynomial function and is preserved as one point on the smoothed curve. One may then add the next successive point on one end and drop the last one on the other end. A similar smoothing operation is performed, and another smoothed data point is added to the calculated curve. A smoothed curve is generated in this manner. The computer filtering saves the unfiltered data, and thus an optimum filter involving a choice of the size of data points and the degree of polynomial may be selected. As digital filtering is symmetric, it does not appreciably distort line shapes, whereas the electronic filter distorts the line shape toward the frequencies measured last. This helps in faithful reproduction of true characteristics of microwave absorptions in any data-reduction procedure based on digital filtering.

Digital averaging of repeated scans of a microwave absorption line is yet another technique used to increase the S:N ratio. Since the spectrometer frequency is quite stable and reproducible, the number of scans over an absorption band can in principle be increased to any desired extent without distorting the shape of the band. The instantaneous average of the signal can be continuously monitored from the oscilloscope display, and the digital averaging can be interrupted when a reasonable S:N ratio is achieved. Some of the features of signal averaging and filtering reported by Gwinn, Luntz, et al. [51] are shown in Figs. 2-8 and 2-9 for the $J = 0 \rightarrow 1$ transition of the

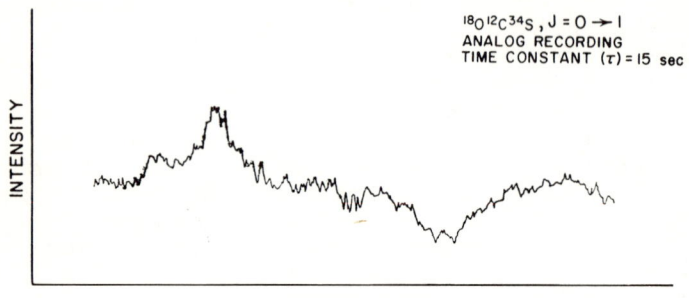

Fig. 2-8. Microwave spectrum of $^{18}O^{12}C^{34}S$, $J = 0 \rightarrow 1$, transition obtained using conventional analog recording with time constant of 15 s.

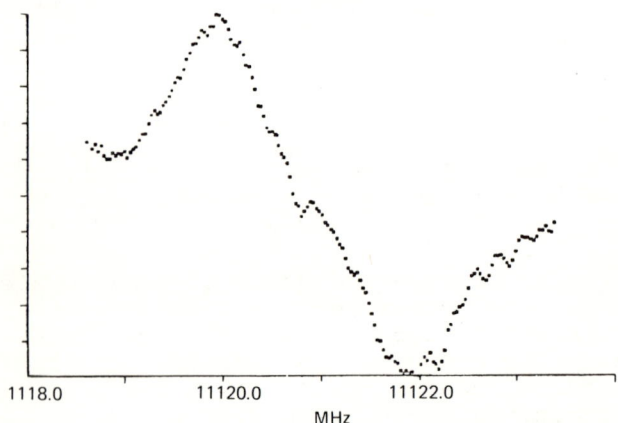

Fig. 2-9. Effect of digital averaging and filtering; microwave spectrum of $^{18}O^{12}C^{34}S$, $J = 0 \rightarrow 1$ transition obtained using integration time of 1 s and 32 scans with smoothing.

$^{18}O^{12}C^{34}S$ molecule. Figure 2-8 represents the signal as detected with a single scan and synchronous detection with a 15-second time constant. Figure 2-9 is a signal from the same transition taken with 32 scans in one second, then processed through a 9-point smoothing routine (requiring an additional 15 s). Note the considerable increase in the S:N ratio.

A modification of technique and the selective application of one or a combination of techniques are required in special situations, such as in search and measurement of extremely weak transitions. One such approach used successfully by White [42] is discussed in detail in Chapter 5. The computer is instructed to modify its scanning procedure as soon as the line is detected. A variable integration time is used at each point. The integration time is increased near the peak frequencies so that when a line is encountered more signal is recorded at certain frequencies. Consequently, the S:N ratio increases continuously as the scan moves near the peak frequency. A remarkable enhancement in signal with less scans is achieved (see Fig. 2-10).

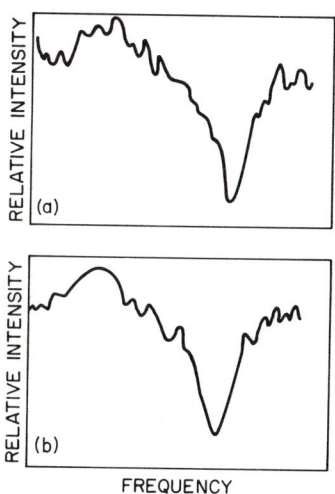

Fig. 2-10. Bottom spectrum (b) was recorded with increased number of scans near peak of band, with net scans (in this case, 3) actually less than that utilized in top spectrum (a) (in that case, 5). A slight enhancement in S:N was observed. (Reproduced with permission of author, W. F. White [42].

Resolution of lines observed by conventional microwave spectrometers is enhanced generally by reducing sample pressure and the microwave power in the cell as well as the modulation frequency used. However, these actions result in lowering of the S:N ratio for the weak transitions. Overlapping of lines makes the situation worse. However, with slight modification of the instrumentation, one can observe and record a second derivative spectrum of possible overlapping lines, thus resolving them to a large extent. The exploitation [42] of the fact that the second derivative of a line shape is much narrower than the line itself is illustrated in Fig. 2-11.

A computer-interfaced microwave spectrometer, operational in the frequency range 8 to 40 GHz using the method of time averaging to achieve improvements in sensitivity, has been reported by Haase, Gegenheimer, et al. [53].

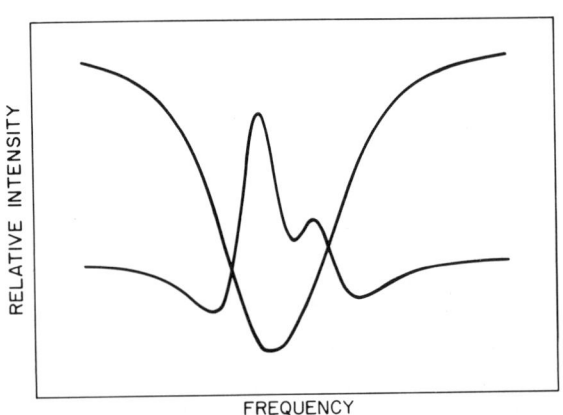

Fig. 2-11. A second derivative spectrum utilized to resolve the broad line (work by White [42]).

2-4. SPECTROMETER AND ABSORPTION CELLS FOR DETECTION AND MONITORING OF CHEMICAL SPECIES OF SHORT LIFE

Analytical applications involving detection, measurement, and monitoring of reaction intermediates may require the use of special absorption cells and the appropriate detection system. Such applications assume particular significance in studying the equilibria or kinetics of chemical reactions conducted in the cell. Additionally, special design features suitable for high-temperature operation may be provided for with proper design of the cell.

Production and maintenance of a high steady-state concentration of reactive chemical species (e.g., free radicals) inside the absorption cell is usually required for monitoring and measurement purposes. The length of a typical absorption cell of a microwave rotational spectrometer makes it difficult to maintain an adequate concentration of the short-lived species in the cell. However, a shorter length for the cell reduces the sensitivity of the spectrometer. For these studies, an absorption cell is chosen to have the maximum length consistent with a reasonable filling factor for the chemical species of short life, or a resonant cavity of high volume:surface ratio is used. There must be some mode of introducing field modulation, either electric or magnetic. An alternative approach may be the use of a simple crystal video-detection at high frequencies. A reasonable spectrometer sensitivity is obtained by performing the measurements at frequencies near or above 100 GHz, since the transitions are more intense at high frequencies.

Microwave rotational spectroscopy of free radicals can be conducted in what is known as a "free-space cell." Such a cell, developed by Kewley, Sastry, et al. [54] is shown in Fig. 2-12. The microwave radiation is fed into the cell through a horn, propagates through the gaseous absorber, and emerges out of the cell through another horn, and then to the detector. Kewley, Sastry, et al. [54] used crystal video techniques for detection and were able to characterize the free radical and the transient molecular species CS and HNCS, respectively, by making measurements on rotational transitions of the chemical species at frequencies up to 250 GHz.

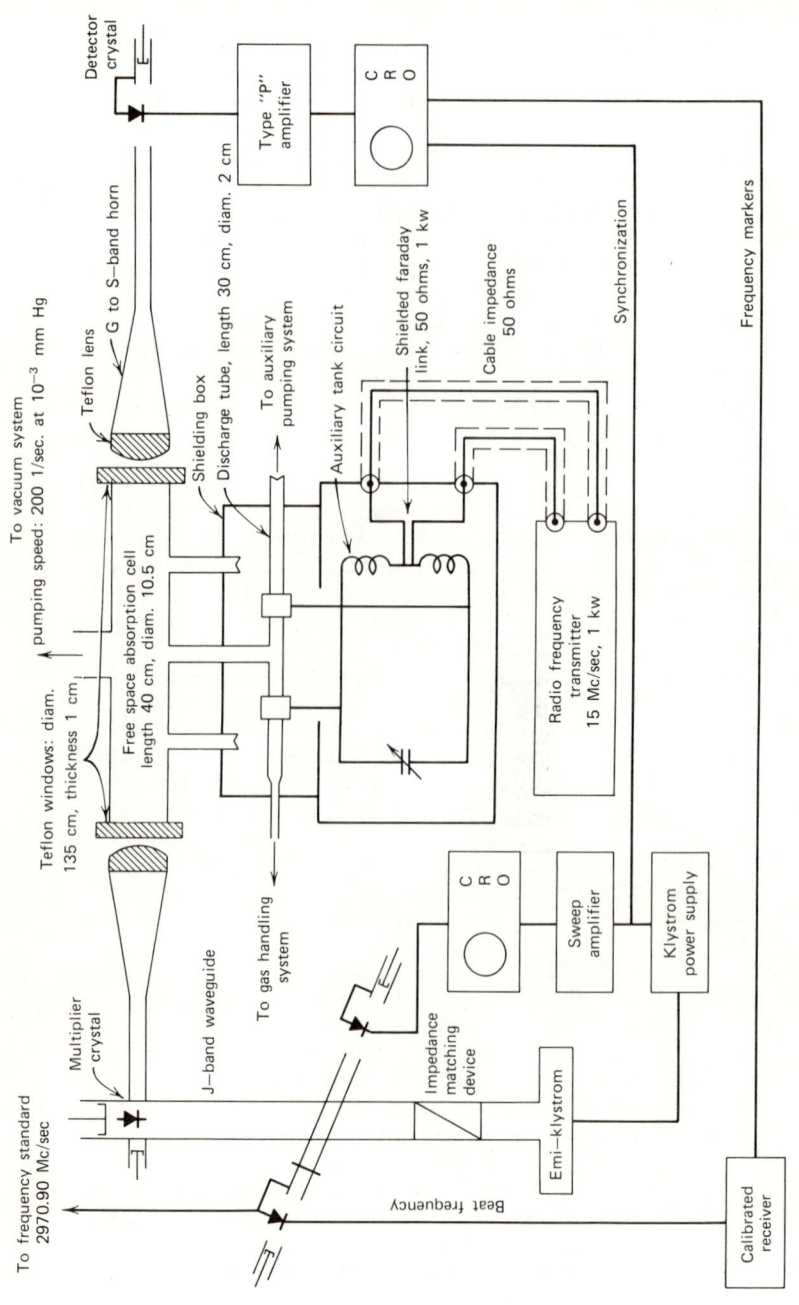

Fig. 2-12. Block diagram of millimeter wave spectrometer and the radio-frequency discharge setup. (With permission of American Institute of Physics.)

The "parallel-plate" cell [15] shown in Fig. 2-13 is typical of Stark modulated free-space absorption cells most commonly used for spectroscopy of short-lived species. The microwave radiation is propagated through the space in the cell between two parallel plates in a manner similar to waveguide propagation. Stark modulation enhances the sensitivity of the system considerably. Examples of measurements from these kind of cells are given later.

A resonant cavity spectrometer, with very high sensitivity in K-band (18 to 26.5 GHz) for gaseous free-radical detection, has been developed by one of the authors [55]. A block diagram of this spectrometer is shown in Fig. 2-14. Resonant cavities, besides having a high volume:surface ratio as an advantage for reactive specie studies, also provide sensitive detection of absorption because of their long equivalent absorption path length. A major

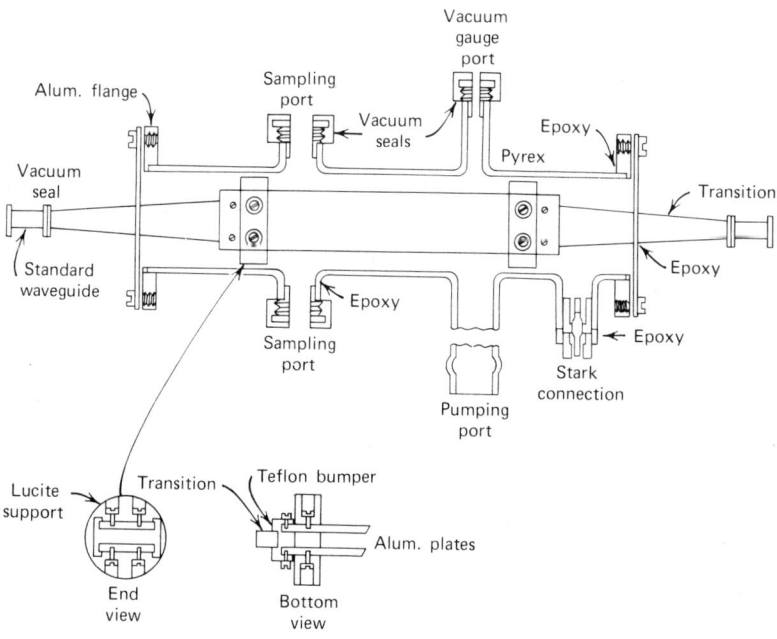

Fig. 2-13. Parallel-plate absorption cell. (With permission of Instrument Society of America.)

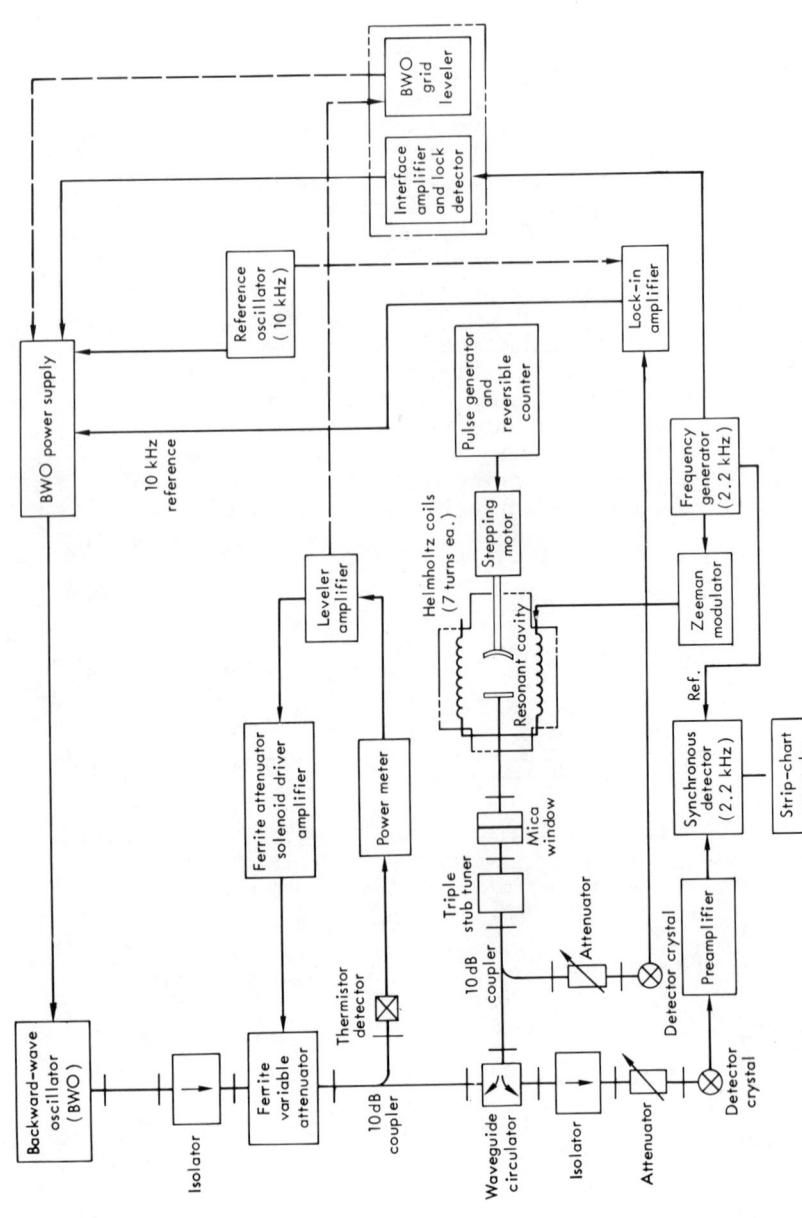

Fig. 2-14. Block diagram of resonant-cavity spectrometer with high sensitivity for detection of gaseous free radicals in K-band.

problem in the application of a resonant cavity absorption cell is that the long path-length advantage occurs only when the oscillator frequency and the cavity resonance frequency are exactly synchronized. For a search spectrometer (i.e., one that can scan a wide bandwidth for unknown absorptions), maintaining such synchronization is difficult.

The spectrometer shown in Fig. 2-14 accomplishes the synchronization of the oscillator to the cavity resonance frequency by an automatic frequency control (AFC) loop. The oscillator, a BWO, is frequency "locked" to the resonant frequency of the cavity by an electronic servo control and follows its every change. Therefore, by mechanically changing the cavity dimensions, and thus its resonant frequency, the system searches a wide bandwidth while maintaining optimum sensitivity.

Magnetic modulation is used in such a spectrometer because the free-radical molecules are paramagnetic (i.e., their energies can be split by the Zeeman effect). Since most stable molecules are diamagnetic, their absorptions will not be modulated by a magnetic field; thus they do not interfere with free-radical absorptions observed with this spectrometer.

The minimum detectable absorption coefficient with this spectrometer, calculated from realistic parameters, is $\sim 2 \times 10^{-11}$ cm^{-1}. The very high sensitivity of this spectrometer has allowed the detection of a very weak spectrum of the NF_2 free radical in the K-band. Individual lines from this species that are expected to have absorption coefficients between 10^{-9} and 10^{-10} were detected by this instrument with a S:N ratio of greater than 5:1.

2-5. SOLID-STATE DIODE MICROWAVE SOURCES AND MICROWAVE RESONANT CAVITIES FOR HIGH-SENSITIVITY SPECTROMETERS

Solid-state microwave oscillators have recently been developed and as yet have not displaced conventional sources for MRS. This is very likely to change soon because such devices have very low noise and are now capable of the same power and tuning range as BWOs. They have the additional advantages of requiring only low-voltage power supplies and have very long lifetimes. Their reliability in other types of microwave circuits is becoming well established.

The most popular solid-state source is called a "Gunn effect diode." It is actually a doped GaAs crystal (not a diode), which has a differential negative resistance region in its I-V characteristic that sustains oscillations at microwave frequencies. The oscillations from a single Gunn diode can be tuned over 10 GHz. This is usually done electronically by applying a tuning signal to either a "varactor" (voltage-variable capacitor) or a YIG sphere (ytterbium iron garnet ferrite material) in close proximity to the Gunn diode in a resonant circuit. Another solid-state source of microwaves is an impatt diode. Such devices can oscillate at higher frequencies than Gunn diodes ($\leq$150 GHz) but are generally considerably noisier.

The Gunn effect diodes usually can generate more than 20 mW of continuous microwave energy at frequencies up to 60 GHz using a 4-V, 0.5-A d.c. power supply. The output frequency of Gunn diodes can be directly controlled by the resonant frequency of the resonator. A resonant cavity is essentially a hollow space enclosed by metallic walls that at resonance provides extremely long path length through multiple reflections and can store considerable amounts of microwave energy. If the quality factor, Q, of the spectrometer is kept high, the microwave radiation has a long path length through the sample gas due to multiple reflections. The block diagram of a Gunn diode microwave-cavity spectrometer is shown in Fig. 2-15. The spectrometer, designed by one of the authors [56], is a combination of battery-powered solid-state Gunn diode, a high-Q resonance cavity, and an interfaced computer for providing automation. A Gunn diode, capable of 40-mW power at the frequency at which the sample gas is expected to absorb, is mounted in a small segment of waveguide that is shorted at one end and coupled to a high-Q Fabry-Perot type resonator at the other end through a small aperture in the flat plate of the resonator. The other plate of the resonator has a spherical concave surface of appropriate curvature to minimize dispersive losses and increase the Q-factor. The small waveguide mount acts as a low Q-resonant circuit to excite the Gunn diode oscillations. The high Q-resonator is then tuned until its resonance frequency equals that of the diode. The output frequency of the diode then locks to the high

Q-resonance. Any change in the frequency of the high
Q-resonant cavity introduces a corresponding identical
change in frequency of the generator. The resonant
frequency of the cavity and thus the oscillator can be
changed mechanically. The computer interfacing makes
it possible to tune the resonant frequency of the
cavity within the limited range of a few GHz. For
analysis of a particular molecule the cavity is tuned

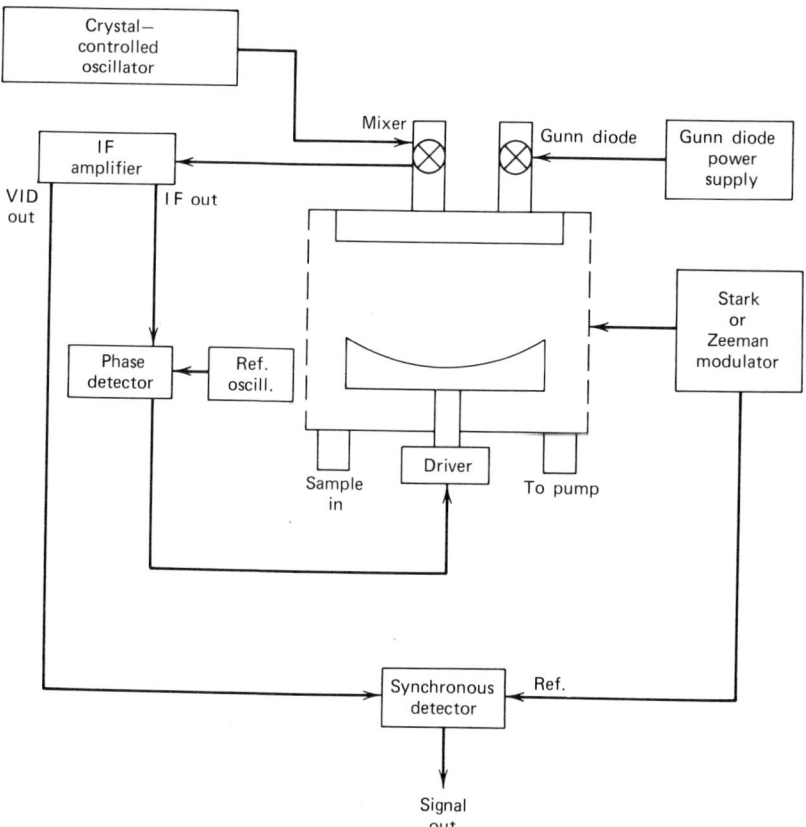

Fig. 2-15. Block diagram of Gunn diode microwave cavity spectrometer. (Work done by Hrubesh, Anderson, et al. [56].)

to a known absorption frequency of the molecule. The
frequency of the resonator is compared with a suitable
harmonic of a stable crystal oscillator. An analog
correction voltage derived from this comparison is
utilized in correcting the physical dimensions of the
cavity and thus further stabilizes the frequency. The
desired frequency is stabilized within 1 part in 10^8
or better. The absorption by the polar molecular
species is detected by Stark modulation and phase-
sensitive detection. Zeeman modulation can be used
for paramagnetic molecules. The practical applications of the system are discussed later. The chief
drawback for such an analysis scheme seems to be the
fact that it has a limited tuning range. The advantages are numerous, such as high sensitivity and portability of the instrumental package to remote places in
the field for monitoring of air pollution.

2-6. HIGH-TEMPERATURE MICROWAVE SPECTROMETER

If the vapor pressure of the sample under study is
too low to provide an optimal concentration of absorbers in the cell at room temperature, high-temperature
cells may be used. Such cells are constructed of
special materials and are designed for high-temperature
operation. Absorption cells working at temperatures
up to 1000°C have been described by Hoeft, Lovas, et
al. [57]. Recognizing the tremendous difficulties
one encounters during the operation of spectrometers
at high temperatures, Hoeft and co-workers have constructed a molecular-beam microwave spectrometer [57].
With this design it was shown that measurements of
rotational spectra can be performed for molecules of
short lifetimes and at effective kinetic temperatures
up to 2000°C.

CHAPTER

3

QUALITATIVE ANALYSIS

Microwave rotational spectroscopy has been used with extraordinary success in the identification and complete characterization of stable molecules, stereoisomers of stable molecules, short-lived reaction intermediates, free radicals, and high-temperature molecules under laboratory conditions.

The frequencies of rotational transitions are determined by the moments of inertia of a molecular rotor; therefore, MRS can be used for specific compound identification. Features such as high resolution, specificity, and precision of MRS are of prime importance for the detection and characterization of any molecular species, provided that spectral transitions of adequate intensity can be observed in each case.

The type of MRS spectrometer selected for qualitative analysis is determined by the kind of chemical species under investigation. For example, a video-type millimeter wave spectrometer in conjunction with a free-space oversized absorption cell is commonly used for measurements on unstable chemical species such as unstable molecules and free radicals. Parallel-plate oversized Stark absorption cells may also be used with the millimeter wave radiation advantageously for the detection of reactive species.

3-1. RESOLUTION AND SENSITIVITY OF MICROWAVE SPECTRA

In microwave rotational spectroscopy, resolution is defined as the peak absorption frequency divided by the full width of the line at the half maximum intensity. Typically, resolution is of the order of approximately 50,000 and with additional care, even 500,000. An example of the high resolution obtainable is shown in Fig. 3-1 for the observed $J = 2 \rightarrow 3$

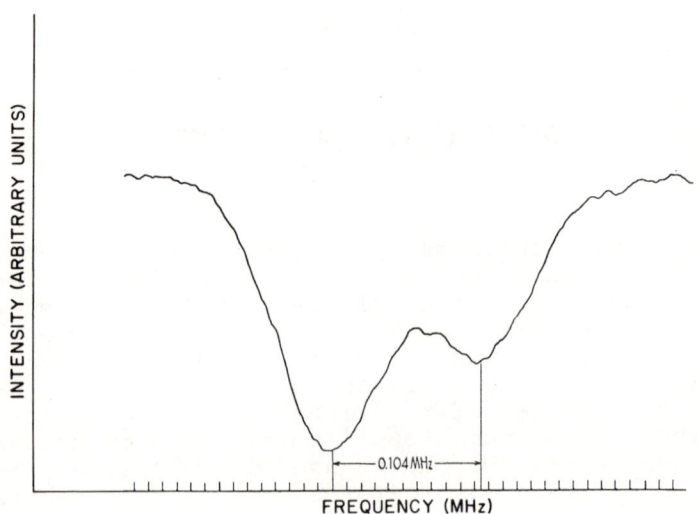

Fig. 3-1. Resolution of J = 2→3 OCS line observed on HP Model 8460A spectrometer with sample pressure less than 0.1 Pa. (With permission of Hewlett-Packard Company.)

transition at 36.4 GHz of the OCS molecule; this has been observed with the HP Model 8460A microwave spectrometer. The pressure of the vapor in the absorption cell is usually kept below 1 Pa to obtain high resolution. However, poor resolution due to accidental overlapping of a transition with one of its neighbors or with Stark lobes of the latter can degrade the sensitivity of measurement.

The absorption coefficient, α_ν, can be defined as the fractional power absorbed by the sample per unit path length. The minimum detectable absorption coefficient, α_{min}, for a spectral line can be defined as one that barely allows observance of the line; that is, the S:N ratio just exceeding 1 is a quantitative measure of the sensitivity of a spectrometer.

As seen from Eq. (1-32), the absorption coefficients are determined by the magnitude of the dipole moment of the molecule, the rotational state of the molecule, and the pressure of the molecules in the absorption cell.

RESOLUTION AND SENSITIVITY OF MICROWAVE SPECTRA

For transitions in the centimeter wavelength region routinely measured by the standard Stark modulated microwave spectrometer, the values of absorption coefficients are found to lie in the range 10^{-5} to 10^{-7} cm^{-1}. If it is assumed that the absorption coefficient of 10^{-9} cm^{-1} is the lower limit of detection for a given spectrometer and that the coefficient of a particular transition is 10^{-4} cm^{-1}, a concentration of 10 p.p.m. or 10^{-5} mole fraction can be determined for that absorbing molecular species. The limit of detection may vary with the particular technique used in measurements. For example, a video-type spectrometer in the millimeter wavelength region has a poor sensitivity, which can be improved to a limiting detection level of 10^{-8} cm^{-1} through time-averaging with the help of a computer. On the other hand, Haase, Gegenheimer, et al. [53] have been able to observe the weak $J = 0 \rightarrow 1$, $F = (3/2) \rightarrow (5/2)$ transition of $^{18}O^{13}C^{32}S$ molecule with an absorption coefficient of 1.3×10^{-10} cm^{-1} (see Fig. 2-7) on a computer-interfaced spectrometer. It may be possible to develop a computer-interfaced microwave rotational spectrometer capable of detecting molecules with absorption coefficients as small as 10^{-12} cm^{-1}. It may be mentioned that a common atmospheric pollutant, NO_2, having an absorption coefficient of 10^{-5} cm^{-1}, could be detected in trace-amount concentration of about 100 p.p.b. by such a spectrometer. On the other hand, for ammonia gas, whose absorption coefficient is approximately 10^{-4} cm^{-1}, 10 p.p.b. could be detectable.

A Gunn diode microwave spectrometer [56] appears promising for the detection of common pollutant gas molecules in specific frequency regions. The minimum detectable absorption coefficient, α_{min} in cm^{-1}, for a resonant-cavity microwave spectrometer is given by the equation

$$\alpha_{min} = \left(\frac{4kTN\Delta f}{P_0}\right)^{1/2} \frac{2\pi}{Q_L \lambda} \text{ cm}^{-1} \qquad (3-1)$$

where

k = the Boltzmann constant;

T = absolute temperature (degrees Kelvin);

N = noise figure for the system;

Δf = frequency bandwidth of the detector (Hz);

P_0 = the power incident on the cavity coupling port, (W);

Q_L = loaded "Q" of the cavity; and

λ = wavelength of the radiation, (cm).

The rotational transition [37] at 28,974.8 MHz of the HCHO molecule, for example, has an absorption coefficient of 6×10^{-5} cm^{-1}. The minimum detectable absorption coefficient, α_{min}, applicable for the cavity-type spectrometer [55] is approximately 2×10^{-11} cm^{-1}. Thus the minimum fractional abundance of HCHO in a gaseous mixture that can be determined by the cavity-type spectrometer is about 0.3 p.p.m.

The high sensitivity claimed for computer-interfaced spectrometers and the cavity-resonance spectrometer is strictly applicable to high-resolution spectroscopy. Low-resolution spectroscopy, on the other hand, can hardly be expected to achieve such high sensitivity.

3-2. SPECIFICITY IN HIGH-RESOLUTION MICROWAVE ROTATIONAL SPECTROSCOPY FOR QUALITATIVE ANALYSIS

High-resolution microwave spectroscopy offers a unique approach for the unambiguous identification and characterization of a gaseous component in a mixture or in a pure state, provided that the rotational absorption spectra can be observed experimentally in each case. A molecule must possess an electric dipole moment if its rotational absorption spectrum is to be observed at all in the vapor state. A unique microwave spectral pattern of both frequencies and relative intensities is associated with any polar molecular species. In other words, two different molecules or even any two isotopic species of the same molecule (e.g., $^{18}O^{13}C^{32}S$ and $^{17}O^{13}C^{32}S$; $^{15}NH_3$ and $^{14}NH_3$) have considerably different rotational spectra, because the moments of inertia or rotational constants are extremely sensitive to isotopic substitution or to slight changes in structural parameters (i.e., bond lengths and angles) of the molecule.

Information on the distribution of frequencies and relative intensities of the rotational transitions observed experimentally can be interpreted to yield an unambiguous identification and can lead to the complete characterization of one or more molecular species in a gaseous sample. However, the observation of many rotational transitions for a typical molecule, for example, in the 26.5 to 40 GHz range, suggests that a great deal of care is needed in the analysis of microwave spectra for identification of a molecular species. A rotational line may be overlapped either by the neighboring lines of other species or by the Stark lobes of its own or other transitions. This may result in distorted line shapes and may also affect observed intensities.

The implications of overlapping of microwave spectral lines for chemical analysis have been examined by Jones and Beers [58]. The number of frequency measurements required for an unambiguous characterization of a molecular species in a gas mixture may depend on the distribution of frequencies and intensities for the rotational transitions of all component gases of the mixture. Jones and Beers [58] have analyzed this problem in a quantitative manner using techniques of statistics. The frequencies of approximately 10,000 rotational transitions arising from 33 polar molecules contained in a hypothetical gas mixture were analyzed with the aid of a digital computer. The variation of average percentage of instances involving overlapping of a rotational transition of one molecule by a line originating from the other 32 molecules was plotted as a function of precision of measurements. The results are shown in Fig. 3-2. It appears that measurement of a single transition frequency for a molecule with a precision of 0.2 MHz corresponds to a theoretically possible overlap in 45% of the cases. If two lines belonging to one molecule are measured even with a precision of 0.2 MHz, overlap is predicted only for 2% of the cases. On the other hand, measurement of three lines with a precision of 0.5 MHz is quite adequate for a positive identification of a particular molecule. The microwave rotational spectrum of a molecule can thus be considered as its fingerprint in a real sense.

A unique distribution of frequencies and intensities for strong and moderately strong transitions predicted for a molecule can be utilized for a complete characterization of the molecular species present in a mixture. The data on relative intensities for many molecules can be readily obtained from NBS linestrength tables, which are calculated from known

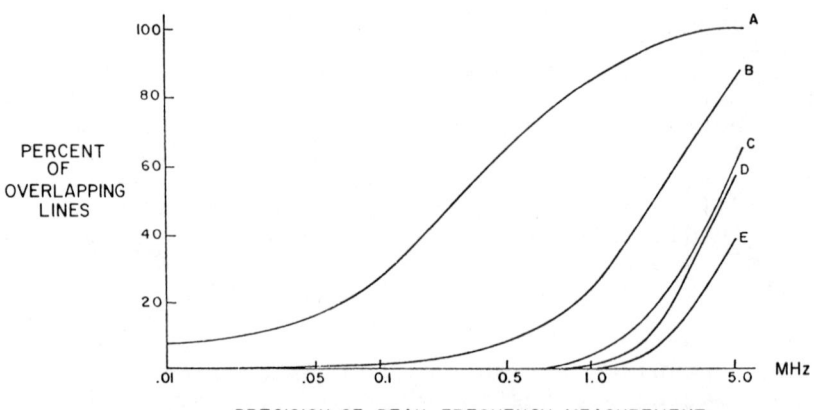

Fig. 3-2. Average percentage of overlapped sets of lines versus precision of measurement for a sample of 33 gases, plotted on a semilogarithmic scale. Each point on curves represents an average of percentage of overlap for 10 molecules. Curve A, one line measured; B, two lines measured; C, three lines measured; D, four lines measured; E, five lines measured. (From Jones and Beers [58]. With permission of American Chemical Society.)

structure and dipole-moment components of the molecule. The National Bureau of Standards has been engaged in compiling a comprehensive set of spectral tables [35] that would list frequencies and line strengths for a large number of molecules. This reference data should be immensely useful for conducting chemical analysis by microwave rotational spectroscopy.

White [59] at the Langley NASA Center has discussed the techniques for identifying the components of gaseous or vapor mixture. The need for reference data [35-37], comprised of frequencies and line strengths for observed transitions of each individual molecular species, is an absolute prerequisite for routine qualitative chemical analysis. Using a computer-controlled HP Model 8460A spectrometer, White has cataloged all lines with absorption coefficients above about 10^{-7}cm^{-1}

for over 100 compounds, including volatile organics, sulfur-and nitrogen-bearing compounds, and some chloro- and fluorocarbons. This catalog is available in several volumes from the NASA [36]. It is the only known comprehensive compendium of microwave absorption lines for particular use by the analytical microwave spectroscopist.

The useful frequency range for chemical analysis is determined by the distribution of frequencies for a large number of polar gases and vapors in the frequency range accessible to a typical microwave spectrometer. An analysis [6] of the distribution curve shown in Fig. 3-3 indicates that the frequencies of the rotational transitions of molecular species to a large degree are observed to lie within ±10 GHz of 25 GHz. However, it is not necessary to use the entire frequency coverage of a spectrometer to detect most molecules since a high percentage of molecules would have reasonably strong rotational lines in the accessible R-band. This is in agreement with the fact that line strength of a transition increases with the square of frequency. Indeed, a careful analysis of the absorption coefficients of transitions in the R-band (see Fig. 3-4) suggests that a large number of molecules absorb strongly in that spectral region.

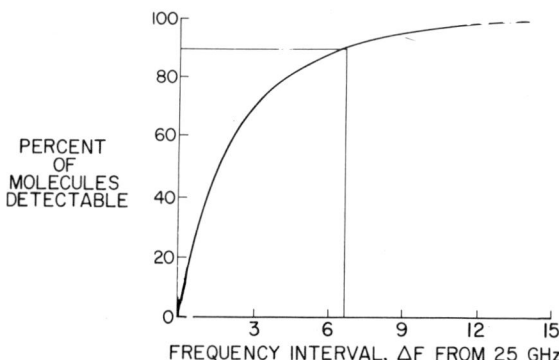

Fig. 3-3. Percentage of molecules detected versus frequency interval from 25 GHz. (With permission of McGraw-Hill.)

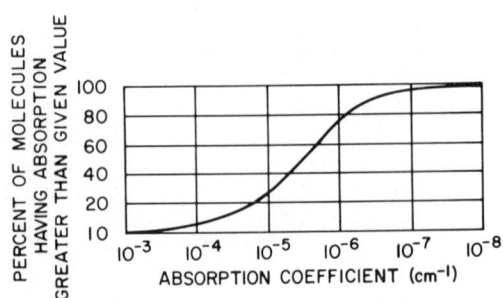

Fig. 3-4. Percent of molecules having absorption greater than given value in R-band. (With permission of McGraw-Hill.)

Experimental conditions such as pressure and temperature of the gaseous substance as well as the level of microwave power in the sample cell are optimized to maximize the observed intensities of the rotational transitions of a particular molecular species. The line width of a transition at half-height, $2\Delta\nu$, of a single component polar gas observed [60] as a function of pressure over the range 0.1 to 25 Pa is shown for a rotational transition of 1-butene in Fig. 3-5. As gas pressure is increased (>0.1 Pa) peak absorption may remain the same, whereas line width generally increases due to enhanced rate of intermolecular collisions.

The half-width at half-height (HWHH) of a rotational transition of a polar molecular species, A, present in a gas mixture along with a second molecular species, B, may be given by collisional cross sections K_{AA} and K_{AB} which depend very much on the dipole moments of the species. Therefore, the HWHH of rotational transitions of a polar species at a specified ($\sim$0.1 Pa) pressure in the presence of a nonpolar diluent gas such as argon will be smaller in comparison to the ones observed for the polar gas sample in the pure state at an equivalent molecular concentration. Measurement on such a binary gaseous mixture at a higher total pressure in the sample cell is expected to provide an improved S:N ratio

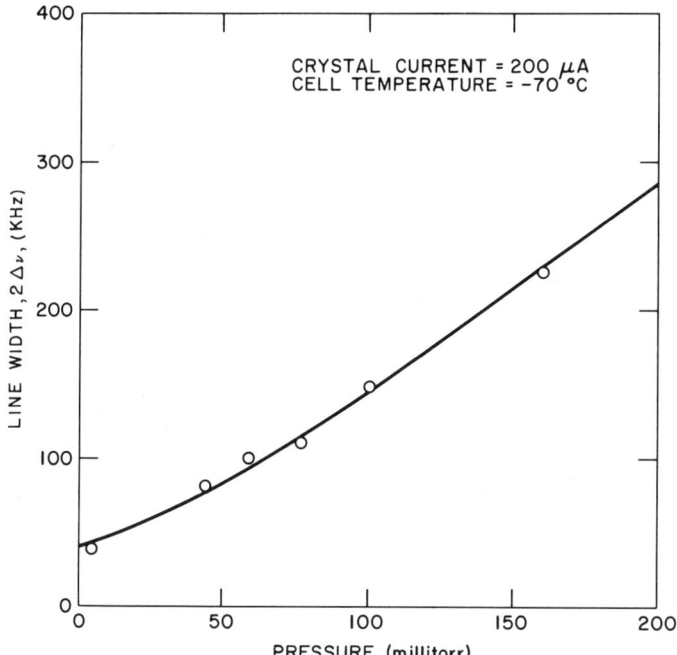

Fig. 3-5. Line width at half height, $2\Delta\nu$, for $2_{12} \rightarrow 2_{13}$ transition of 1-butene as a function sample pressure in Stark cell at -70°C. (With permission from Society De Chimie Physique.)

for the rotational transitions of the polar molecular species present at a relatively low concentration.

Observation of rotational transitions under conditions that promote high resolution (e.g., low sample pressure and fully modulated line shapes), is fully capable of unambiguous detection and characterization of a molecular component in a mixture. However, the time consumed in scanning the entire spectrum in the R-band to obtain high-resolution data with a non-automated spectrometer may be long. The scan time as well as the frequency range scanned can be varied over a wide range with the modern microwave spectrometer.

The fastest scan rate that can be used in a given case depends on the strength and width of the line or lines under observation, since these parameters determine the detector time constant. Also, these absorption parameters are directly controlled by the sample composition and pressure. High pressures (≥ 13 Pa) naturally permit use of higher scan rates, but only at the cost of resolution. At the higher pressures, there is also a possibility that arcing in the sample cell may damage the sample and/or the Stark electrodes. Use of relatively low Stark modulating fields enables one to use higher pressures in the cell. This appears to generate a simple spectral pattern, since many of the lines are not observed as they are severely undermodulated.

A definite advantage in a microwave rotational spectrometer is the ability to program the excitation frequency to exact coincidence with absorption peaks, if those peak frequencies are known a priori. Then rather than scanning over the absorption line, the scan time can be used to signal average while sitting on the peak. A spectrometer based on this method is described in Chapter 5. It is capable of carrying out a complete qualitative search and semiquantitative analysis for over 100 different compounds in less than 3 min.

3-3. APPLICATIONS OF HIGH-RESOLUTION MICROWAVE SPECTROSCOPY IN IDENTIFICATION AND CHARACTERIZATION OF MOLECULAR SPECIES, REACTION INTERMEDIATES, AND FREE RADICALS

The microwave rotational spectrum of a molecule is its fingerprint. We have discussed before in detail how a determination of frequencies of only a few medium and strong rotational transitions would be sufficient for identification of a molecule. Furthermore, additional information about relative intensities of the same transitions would add to the confidence level of the determination.

Various interesting molecular species have been completely characterized by MRS. The experimental conditions found favorable for generation of these molecules, the manner in which they have been detected, and the cited reference for the investigations appear in Table 3-1.

Short-lived unstable species such as free radicals and reaction intermediates have been characterized by observing the rotational spectra in the millimeter or submillimeter wave regions. Numerous studies of recent origin have been directed toward the analysis of intermediates formed when reactions are conducted under energetic (e.g., pyrolysis, radiofrequency or electrical discharge, microwave discharge) conditions or in the presence of a catalyst. Studies on free radicals and other short-lived species have been carried out in the microwave absorption cells shown in Fig. 2-12 and 2-13.

The transient molecular fragment CS was produced [54] by radiofrequency discharge in tubes attached to the bottom of the free-space cylindrical absorption cell shown in Fig. 2-12. The flow rate of CS_2 gas into the continuously pumped sample cell was adjusted so as to produce a steady-state concentration of CS species in the cell. The carbon monosulfide species $C^{32}S$ and $C^{34}S$ were characterized through measurement of rotational transitions in the frequency region 45 to 250 GHz.

The parallel-plate absorption cell shown in Fig. 2-13 has been used at room temperature for detection of reaction products from the pyrolysis of ethylamine [15]. The pyrolysis system consisted of a thin quartz tube heated over a length of 4 cm in a tube furnace. The products of reaction were pumped continuously through the parallel-plate absorption cell. A short-lived species, $CH_2=NH$ and others such as NH_3, HCN, CH_3CN, $CH_2=CHNH_2$ (vinylamine), cis and trans c-methylmethylenimine (CH_3CH-NH) were identified from their rotational spectra.

Species having lifetimes of milliseconds have been readily detected by the MRS technique. Recently studied transient species, which include radicals and short-lived molecules produced by pyrolysis, detected and characterized by microwave rotational spectroscopy are given in Table 3-2.

The reaction products of pyrolysis [15] or catalyzed deuterium exchange [43,61] can be identified and completely characterized. As is shown in a later chapter, this is only the first step in studying a chemical reaction quantitatively. Some of the interesting chemical reaction products detected are shown in Table 3-3.

TABLE 3-1. IDENTIFICATION AND CHARACTERIZATION OF MOLECULAR SPECIES BY HIGH-RESOLUTION MICROWAVE SPECTROSCOPY

Molecule	Experimental Conditions for Measurement	Rotational Transitions Measured	Frequencies /Frequency Range (MHz)	Relative Intensity	Reference No.
$(HF)_6$	Total pressure of 133 Pa at 203 K; 4.1% $(HF)_6$ in a mixture of $(HF)_x$ components where x = 1,2,3,4,5,7	a-type and b-type	9,000 to 24,000	—	62
$(HF)_7$	Total pressure of 133 Pa at 203 K; 3.5% $(HF)_7$ in a mixture of $(HF)_x$ components where X = 1,2,3,4,5,6	a-type and b-type	9,000 to 24,000	—	62
$H_2^{16}O\cdots HF$	One:one mixture of H_2O and HF at total pressure of ~100 Pa	$0_{00} \to 1_{01}$ $1_{10} \to 2_{11}$ $1_{01} \to 2_{02}$ $1_{11} \to 2_{12}$	14,403 28,920.7 28,804.7 28,673.4	—	63
$^{16}O^{18}O$	Gas-phase magnetic dipole allowed transitions observed	$n(J) = 0(1) \to 2(1)$	233,946.178	—	64
trans CH[:O]SH (H-C(=O)-S-H)	Spectra recorded at room temperature on HP Model 8460A spectrometer	a-type and b-type	—	—	65
cis CH[:O]SH (H-C(=O)-S-H)	Spectra recorded at room temperature on HP Model 8460A spectrometer	a-type and b-type	—	—	65

Compound	Description	Notes	Ref
CH_3CHS Thioacet- aldehyde	Pyrolysis of S-triathane; lifetime (0.4 Pa pressure in gold-plated cell): 10 s	—	66
$(CH_3)_2CS$ Thioace- tone	Pyrolysis of hexamethyl S-trithiane; lifetime (0.4 Pa pressure in gold-plated cell): >10 s	—	67
$ClNO_3$	Compound prepared by reacting ClF with HNO_3; stainless-steel wave-guide and Stark septum used; pressure of vapor during measurement over 13 Pa.	a-type and b-type R-branch lines measured; $^{35}ClNO_3$ and $^{37}ClNO_3$	68
$O=C=C=C=S$	Sample prepared by reaction of C_3O_2 with P_4S_{10}	b-type doubling observed for ν_5, ν_6, and ν_7	69
MnO_3F	Spectra recorded on Model HP8460A spectrometer with sample allowed to flow through the sample cell in absence of moisture and air	ground $\nu_3 = 1$, $\nu_5 = 1$, and $\nu_6 = 1$ states observed along with some quadrupole hyperfine structure	70
$(CH_3)_3NBF_3$	Reaction of $(CH_3)_3N$ with B_2H_6 in gas phase in sample cell	—	71
$(CH_3)_3PBH_3$	Reaction of $(CH_3)_3P$ with B_2H_6 in sample cell	—	71
$CH_3PH_2BH_3$	Reaction of CH_3PH_2 with B_2H_6 in sample cell	—	71

TABLE 3-1 (Continued)

Molecule	Experimental Conditions for Measurement	Rotational Transitions Measured	Frequencies /Frequency Range (MHz)	Relative Intensity	Reference No.
HNC	Mixtures of C_2N_2 and H_2, C_2N_2 and C_2H_2, as well as N_2 and C_2H_2 irradiated with d.c. glow discharge using d.c. glow-discharge microwave spectrometer				72
HCN···HCN		$HC^{14}N\cdots HC^{14}N$, $HC^{15}N\cdots HC^{15}N$, and DCN···DCN species were assigned and molecular constants obtained			73

TABLE 3-2. TRANSIENT SPECIES DETECTED BY MICROWAVE SPECTROSCOPY

Radical or Short-lived Molecular Species	Reference
HO_2	74
OH	75
$SO\,^3\Sigma$	76
$SO\,^1\Delta$	77
S_2O	78
S_2O_2	79
SF	80
SF_2	81
NS	82
$C\ell O$	83
BrO	84
IO	85
BF	86
NCO	87
HNO	88
HCO	89
SiF_2	90
GeF_2	91
NF_2	97

TABLE 3-3. PRODUCTS OF PYROLYSIS OF STABLE MOLECULES AND CATALYZED CHEMICAL REACTIONS DETECTED BY MICROWAVE SPECTROSCOPY

Molecules	Reference
Products of pyrolysis	
$CH_2=NH$	15
$CH_2=CH-NH_2$	
$CH_3CH=NH$	
Products of catalyzed deuterium exchange	
1-Butene and d_1-derivatives	61
Propene-d_1	43
Propene-d_0	43

3-4. QUALITATIVE ANALYSIS ON COMPUTER-CONTROLLED MICROWAVE SPECTROMETER

A molecular species is identified by matching the observed frequencies of the rotational transitions with those previously determined for a pure sample of the particular compound. As relative intensities of transitions are also characteristic and unique for an individual molecule, they may be helpful in confirming the identity of the molecule. A list of strong rotational transitions for a suspected molecular species in a gaseous mixture can be obtained from published [35-37] or calculated line-strength data or from experimentally observed relative intensity data for the molecule. The experimentalist can then save time by restricting his or her efforts to the measurements of frequencies and relative intensities of strong transitions only.

A computer-interfaced automated microwave spectrometer is quite appropriate for conducting routine qualitative analysis of gas mixtures in an automated manner. The essential components of a successful program plan to achieve this goal are discussed in the paragraphs that follow.

Reference data on frequencies and intensity parameters for medium to strong rotational transitions for a large number of molecules should be stored in or be available to the computer interfaced to the microwave rotational spectrometer. The total data on a typical molecule do not have to be extensive and may consist of frequencies and relative intensities or line strengths for 20 to 30 medium to strong transitions. The "Joint Committee on Atomic and Molecular Physical Data," which included representatives from NBS, NASA, and universities, have recently recognized the need for assembling comprehensive spectral data and making them available to those engaged in chemical analysis work. This task is currently being pursued by the spectroscopy section of the National Bureau of Standards in Washington, D.C. The spectral tables [35] or computer-stored data would normally provide the frequencies and a measure of relative intensities. The intensity parameter $-10 \log \alpha_{\nu_0}$ where α_{ν_0} is the Beers law unsaturated absorption coefficient, has been adopted as the most convenient way to handle coefficients that may vary over many orders of magnitude. Furthermore, this parameter can be easily converted to the decibel scale, on which a microwave spectrometer is usually calibrated.

Control of all major spectrometer functions such as search and measurement of spectral lines, collection of experimental data, and reduction of data may be performed in a routine manner by interfacing a suitable digital computer to the HP Model 8460A microwave spectrometer. White [59] has written a comprehensive test computer program that would perform all the tasks mentioned earlier. A scheme suitable for conducting qualitative chemical analysis is briefly described in the paragraphs that follow.

The computer programs the spectrometer to the strongest line of the first molecule and also uses the stored intensity data to decide on an initial gain setting. It looks for the signal, and if none is observed the gain is increased and the signal repeatedly monitored until the line is detected or the limiting gain for the spectrometer is reached. In case a line is not detected, search is started for the strongest lines of other molecules using the available reference data. The program looks for a frequency match within a certain error limit chosen beforehand by the analyst. A printout of status report on possible matches as well as the upper limits of concentration is available after search has been completed for the strongest transition of all the molecules. At the same time the computer is allowed to make a decision to measure precisely the frequency as well as relative intensity in the event that match between any observed and calculated transition is excellent, that is, within a certain preselected error limit. The computer will now proceed on to the task of search and measurement (if indicated) of the second, third, and fourth strongest lines for each molecule from the stored data bank. The final status report printout may, for example, show a positive identification of a component or components of a gaseous mixture unambiguously. A typical spectrum recorded by White [59] for a mixture of 1-propanol, methylene chloride, methyl ethyl ketone, and 2-propanol vapors is reproduced in Fig. 3-6. The limited number of rotational lines identified by a process such as that outlined

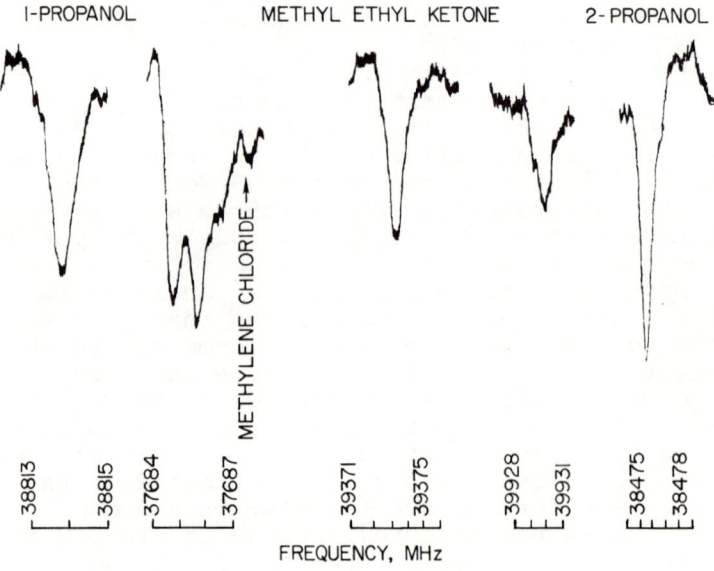

Fig. 3-6. High-resolution spectra of gas mixture. (With permission of American Chemical Society.)

QUALITATIVE ANALYSIS ON COMPUTER-CONTROLLED MS   77

in this paragraph may be quite adequate for a definitive identification. Procedures such as this may be developed further for routine operation. A great deal of experimental work is needed to complete such a task.

3-5. LOW-RESOLUTION MICROWAVE-BAND SPECTROSCOPY FOR QUALITATIVE ANALYSIS

The low-resolution microwave band spectra (LMWBS) of large near-symmetric top polar molecules in the vapor phase, obtained by using typically fast scan rates of 2 to 10 MHz/s and large detector time constants of 0.2 to 1 s on automated commercial spectrometers may be utilized for identification as well as characterization of molecular species and in certain cases molecular conformers. Some large polar organic molecules are near prolate symmetric tops and may give rise to such band spectra.

Different degrees of resolution achieved for the CH_2Cl_2 spectrum recorded by White [59] illustrates the remarkable dependence of resolution on scan rate and the detector time constant. Fig. 3-7 shows a low-resolution spectrum of CH_2Cl_2 at 266 Pa pressure in just about one minute. This spectrum is obtained with a heavily damped detector. The portion of the spectrum indicated by markers in Fig. 3-7 can be scanned in one hour to yield a spectrum of improved resolution (at 20 Pa pressure) as shown in Fig. 3-8. Further improvement in resolution of the band marked in Fig. 3-8 was achieved (at 5 Pa pressure) by scanning over the small frequency range for about 15 min and is shown for comparison purpose in Fig. 3-9. The single strong line shown in Fig. 3-9 is observed as a triplet by scanning at a further reduced rate. A highly resolved spectrum (see Fig. 3-10) has now been achieved. Accurate measurement of frequency as well as relative intensity of all the 1200 or so cataloged lines in the R-band of CH_2Cl_2 would take a day or so. Low-resolution data obtained in the manner illustrated earlier may be sufficient for identification of molecules in a mixture in certain cases. The theory and practice of low-resolution microwave spectroscopy is now presented so that the reader can comprehend the full potential of this technique.

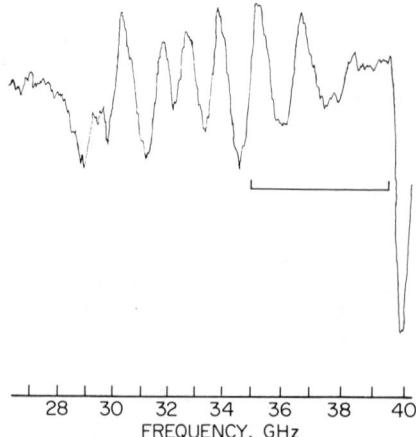

Fig. 3-7. Low-resolution spectrum of CH_2Cl_2 at 266 Pa pressure. (With permission of American Chemical Society.)

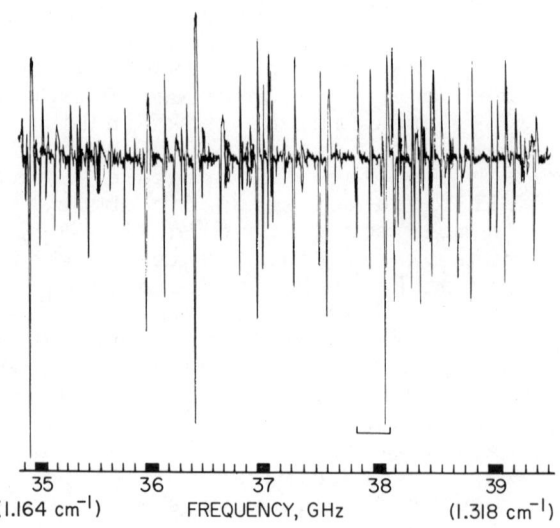

Fig. 3-8. Spectrum of CH_2Cl_2 at 20 Pa pressure. (With permission of American Chemical Society.)

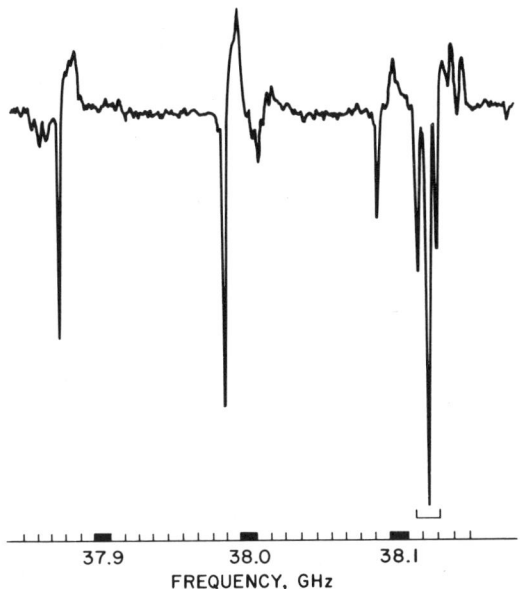

Fig. 3-9. Portion of CH_2Cl_2 spectrum at 5 Pa pressure. (With permission of American Chemical Society.)

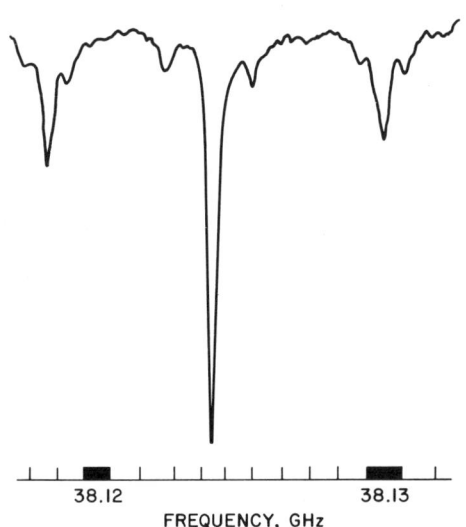

Fig. 3-10. Diagram showing $11_{0,11} \rightarrow 11_{1,10}$ transition of CH_2Cl_2 at 1.3 Pa pressure. (With permission of American Chemical Society.)

The principles of LMWBS have been discussed by a number of authors [14,92-96] of published articles in the field. Many large organic molecules are near symmetric prolate tops. The energy, $E_{J,K_{-1}}$, of the rotational energy levels of a prolate symmetric rotor can be expressed by the equation

$$E_{J,K_{-1}} = (A - \frac{B+C}{2})K_{-1}^2 + (\frac{B+C}{2})J(J+1) \qquad (3-2)$$

The rotational constants (A, B, and C) and the quantum numbers (J, K_{-1}, and K_{+1}) have been defined earlier in Section 1-2. The component, μ_a, of dipole moment along the a-axis in the case of a prolate top is responsible for the a-type transitions with absorption selection rules, $\Delta K_{-1} = 0$ and $\Delta J = 1$. In the case of a near-prolate top with κ (Rays's asymmetry parameter defined earlier) in the range -0.70 to -1.0, the high-resolution spectra would generally consist of clusters of resolved spectral lines. Under low-resolution conditions the clusters of these lines as well as their vibrational satellites are observed as broad bands. The peak frequencies of these bands, ν_{J_a}, can be fitted approximately to an equation of the form

$$\nu_{J_a} = (B+C)(J+1) \qquad (3-3)$$

where J is an integer. Molecules with κ in the range $-0.70 > K > -1.0$ can exhibit such band spectra (hereinafter called Type I), especially for high-J rotational transitions. The average spacing between the Type I bands [Eq. (3-3)] equals B + C, which is very close to the sum of corresponding ground-state rotational constants, $B_0 + C_0$. The peak frequencies of the less commonly observed (Type II) b-type bands, as a result of superposition of b-type high-resolution transitions (selection rules: $\Delta K_{-1} = 1$, $\Delta J = 0$), can be represented by an equation of the form

$$\nu_{J_b} = (2A - B - C)(K_{-1} + 1/2) \qquad (3-4)$$

A third type of equally spaced bands that originate from superposition of clusters of a-type R-branch lines as well as accompanying vibrational satellites were predicted by Ikeda, Curl, et al. [94]. The peak frequency of such bands is expressed by the equation

$$\nu_{J_a} \overset{\sim}{\sim} 2C\,(J + 1/2) + 1/2(A + B) \tag{3-5}$$

Therefore, the spectra consist of bands with frequency separation of ~2C. These new type of bands are hereinafter described as Type C bands.

A molecular species of appropriate size and asymmetry may exhibit band spectra of I, and II, and C types described earlier (see Fig. 3-11). The same molecular species may exist in more than one stable stereoisomer, with each conformer responsible for a series of spectral bands. One may then observe multiple series of bands by the low-resolution technique. The band spectral patterns can be analyzed to identify and characterize a molecule as well as their conformers in favorable circumstances.

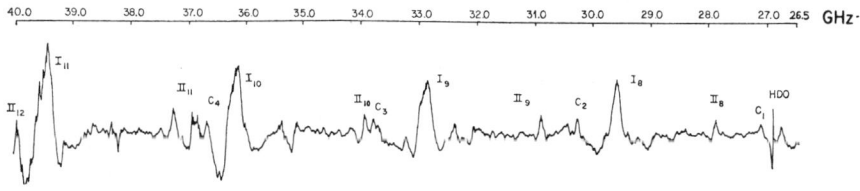

Fig. 3-11. Low resolution spectrum of 1,1,difluorocyclohexane from 26.5 to 40.0 GHz. Four strong Type I bands, five weak Type II bands, and several other weak bands due to Type C transitions are shown. (With permission of Academic Press.)

3-6. APPLICATIONS OF LMWBS IN IDENTIFICATION AND CHARACTERIZATION OF RELATIVELY LARGE ORGANIC MOLECULES

The low-resolution band spectra (LMWBS) of a molecular species in pure state or of more than one pure component in a gaseous or vapor-mixture can be analyzed in a qualitative manner to yield information as to the identity of the conformers of pure components themselves. The LMWBS [95] of p-bromoethylbenzene (^{81}Br and ^{79}Br) and exo- and endo-2-norbornane-carbonitrile are shown in Fig. 3-12 to illustrate the general features of such spectra.

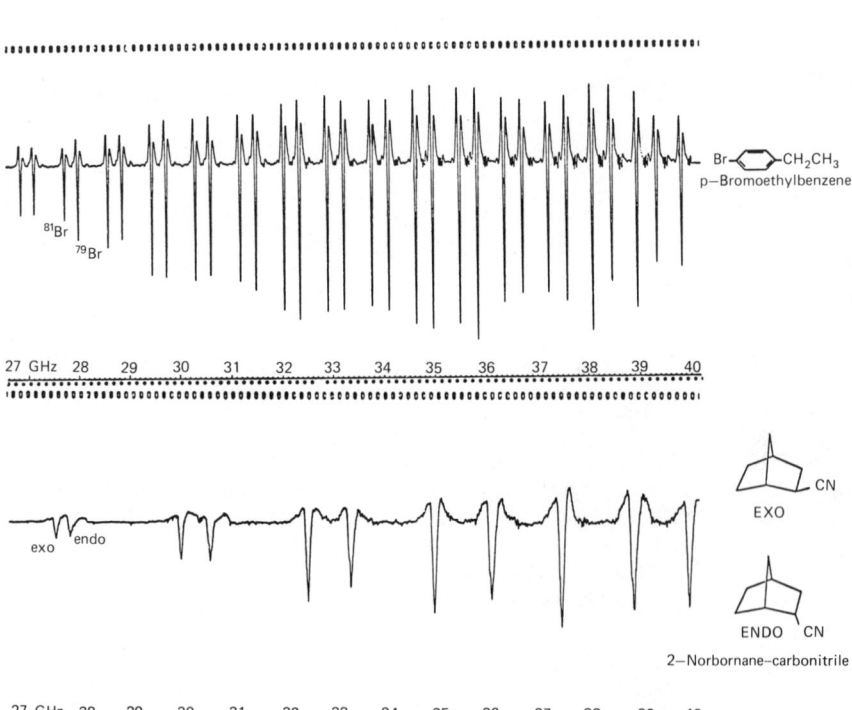

Fig. 3-12. Low-resolution spectra of p-bromoethyl benzene and 2-norbornane-carbonitrile. (With permission of Hewlett Packard Company.)

The total number of band series of one type observed for a particular molecule would, in absence of isotopic species, directly give the number of conformers. In the case of the presence of isotopic species in the molecule, the number is directly proportional to the number of isotopic variants (see Fig. 3-12). Considering that the differences $\Delta(B_0 + C_0)$ or $\Delta(2A_0 - B_0 - C_0)$ for the various conformers of a molecule can be calculated accurately using a set of good assumed structural parameters, a comparison of these calculated values with those observed experimentally may then be utilized to identify conformers in a direct manner. Occasionally additional information available from other investigations such as proton magnetic resonance (PMR) and infrared (IR) spectroscopy, can be utilized to confirm such assignments. The values of $B + C$ and $2A - B - C$ obtained from an analysis of Types I and II band spectra and its comparison with calculated values in a few instances may directly reveal the identity of the molecule. For example, the R-band spectra [96] of 2,2,2-trifluoroethylvinyl ether (F_3-EVE) consist of six strong Type I and two broad Type II bands. Analysis [96] of the spectral data of the molecule (Fig. 3-13) is presented in Table 3-4. The Type I bands are spaced at 2181.9 ± 0.3 MHz intervals, thus yielding an average value for $(B + C)$. Using an estimated value of -0.97 (calculated from assumed structural parameter shown in Fig. 3-14) for the asymmetry parameter, κ, in the empirical equation [96]

$$\frac{B + C}{B_0 + C_0} = 1 + (0.025 \pm 0.005)(1 + \kappa) \qquad (3\text{-}6)$$

and an estimated value of 2180.3 ± 1.1 MHz for $(B_0 + C_0)$, an approximate value is obtained for $(B + C)$. An assumption is made that the effective ground-state constant $(2A_0 - B_0 - C_0)$ can be equated to the observed effective constant $(2A - B - C)$, obtained from the Type II data. Thus the values of A_0 and $(B_0 + C_0)$ constants are determined. This analysis shows that in certain situations an unambiguous identification and characterization of a molecular species and/or conformer is possible with LMWBS techniques.

TABLE 3-4. LOW-RESOLUTION MICROWAVE BAND SPECTRAL DATA OF 2,2,2-TRIFLUOROETHYLVINYL ETHER

Type I Bands	$J + 1$	Observed ν (MHz)	$B + C$ (MHz)
	13	28,365	2181.9
	14	30,550	2182.1
	15	32,730	2182.0
	16	34,910	2181.9
	17	37,100	2182.4
	18	39,265	2181.4
Average			2181.9 ± 3
Calculated[a]	κ		−0.97 ± 0.01
Estimated	$B_0 + C_0$		2180.3 ± 1.1

Type II Bands	$K_{-1} + 1/2$	Observed ν (MHz)	$2A - B - C$ (MHz)
	3.5	23,370	6677
	4.5	29,950	6656
	5.5	36,655	6665
Average			6666 ± 12
A rotational constant			4424 ± 7

[a]For rotational isomer shown in Fig. 3-14.

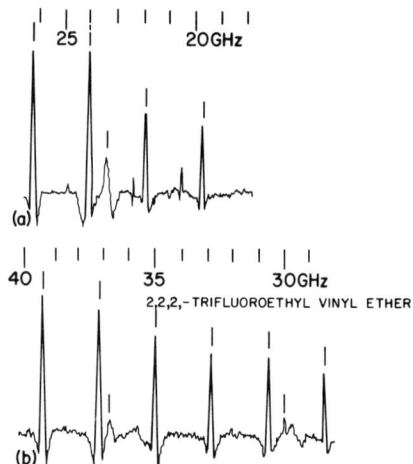

Fig. 3-13. Low-resolution spectra of 2,2,2-trifluoroethylvinyl ether. (With permission of The American Institute of Physics.)

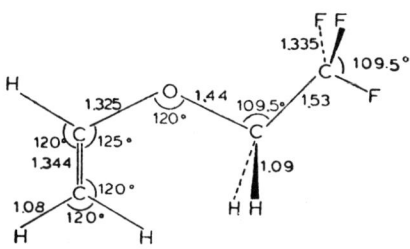

Fig. 3-14. Structural parameters of 2,2,2-trifluoroethylvinyl ether.

CHAPTER

4

QUANTITATIVE ANALYSIS BY MICROWAVE SPECTROSCOPY

For analytical methods utilizing low power level, the microwave power level is kept low in most analytical determinations so that the populations in the lower and upper rotational states are in thermodynamic equilibrium (see Fig. 1-2, Case I) and are controlled by the Boltzmann distribution law. Furthermore, the pressure of the same gas or vapor during measurement is kept at a low level (~ 1 Pa) in the absorption cell, which is consistent with a line width, $2(\Delta\nu)$, of the order of a few megahertz, to ensure the highest possible resolution. Equation (1-28) for the absorption coefficient, α_ν, for a microwave transition may be rewritten as

$$\alpha_\nu = B(N \cdot F_{J,K_-,K_+}) \nu^2 |\mu_{ij}|^2 [\frac{\Delta\nu}{(\nu-\nu_0)^2 + \Delta\nu^2}],$$

where (4-1)

$$B = \frac{8\pi^2}{3ckT(2J+1)},$$

μ_{ij} = dipole moment matrix elements for the transition

$J, K_-, K_+ \rightarrow J', K'_-, K'_+$.

The remaining symbols are the same as those defined in Section 1-5. The factors outlined in the list that follows should be considered in choosing rotational transitions and other experimental conditions for intensity measurements:

1. Molecules with large electric dipole moments give rise to rotational transitions characterized by a larger absorption coefficient α_ν.

2. The intensity of a rotational transition increases as a function of ν^2.

3. The absorption coefficient, α_ν, of a rotational transition of a molecular species at a constant temperature is directly proportional to the number of molecules per cm^3 in the lower rotational state J, K_-K_+.

4. The intensity of a rotational transition increases with decrease in temperature if other parameters are kept constant. Pressure-broadened line widths decrease with temperature.

Line-broadening effects are important considerations in the experimental determination of absorption coefficients. Doppler and modulation effects as well as the collisions of molecules with the walls of the cell contribute to a Gaussian line shape for the rotational transition observed at sample pressures less than 0.1 Pa. However, a Lorentzian line shape is observed for a transition with higher sample pressures as a result of intermolecular collision effects, which also give rise to line width.

Determinations of intensities of microwave absorption lines in analytical applications can involve the measurement of peak height alone, peak height and line width, or the integrated absorption coefficient. Usually, the pressure of a gaseous sample in the cell is maintained near 1 to 13 Pa during the measurement since resolution becomes poorer at higher pressures due to increased line widths.

4-1. PEAK INTENSITY OF ROTATIONAL TRANSITIONS

Measurement of peak intensities of rotational transitions (i.e., $\nu = \nu_0$) constitutes one of the most common approaches to quantitative chemical analysis by microwave spectroscopy. A modified version of Eq. (1-33), such as the relation

$$\alpha_{\nu_0} = \frac{K'p}{\Delta\nu}, \qquad (4-2)$$

is applicable to a gas or vapor in the microwave absorption cell at a given temperature. The term K' is a constant at a fixed temperature and p is the sample pressure. One half the line width at half its maximum height (HWHH), $\Delta\nu$, is proportional to the pressure of the gas in the pressure range 1 to 133 Pa, where collisions are known to dominate the line width. Therefore, the peak intensity remains constant for the same range of sample pressure as shown in Fig. 11 because the pressure dependence in the numerator of Eq. (4-2) is canceled by the pressure dependence of $\Delta\nu$ in the denominator. The peak intensity of rotational transitions of a component gas in a mixture is proportional to the partial pressure of the gas only in special cases.

The interaction between molecules of different kinds in the mixture must be considered so as to lay the proper basis for understanding the relationship between partial pressure of a gaseous component and the observed peak intensity. For example, the collisional broadening effect of the molecules of a substance X in the gaseous mixture on the molecules of substance A may be different from the corresponding effects of the molecules of A among themselves. Thus the spectral line width changes with the composition of the mixture at the same total pressure. The peak absorption coefficient of a transition would be proportional to the partial pressure of the component gas in the mixture if the line width of a transition of a component gas were directly proportional to the total pressure of the mixture. This is usually the case, for example, when the component of interest is dilute within a major component (e.g., trace gases in atmosphere). The observed peak intensities of an ammonia spectral line in a mixture of argon and ammonia gas constitutes a clear example of failure of Beer's law due to differing line-broadening collision parameters of the individual molecular species. The peak height of the ammonia line decreases to only about half of the value in pure state when the gas is diluted with argon so that the ammonia content in the resultant mixture is 5%. The total pressures in the absorption cell are kept at the same value during the experiment. The collisional cross-section of one ammonia molecule against another is several times greater than the corresponding cross-section of the ammonia molecule for an argon atom (molecule) because of the differences in dipole moment of the two species. Therefore, an acceptable analytical procedure based on measurement of peak intensities should take into consideration the different collision-related broadening factors of all molecular species in a mixture. The full spectral width at half maximum height, $2(\Delta\nu_A)$, of a rotational transition of a component gas, A, in a mixture may be expressed by an empirical equation [98] of the form

$$2(\Delta\nu_A) = K_{AA}p_A + K_{AB}p_B \cdots + K_{AX}p_X \cdots \qquad (4\text{-}3)$$

where the coefficient, K_{AX} (X = B,C,D, etc.), is a measure of the collision-broadening factor or efficiency of molecules of X on those of A and p_X is the partial pressure of X. Comparison of Eqs. (4-3) and (4-2) shows that the peak-intensity coefficient, $\alpha_{\nu_{0_A}}$, of rotational transitions of A would not be proportional to its mole fraction in the mixture unless all K values were equal.

The peak intensity of the line for component A in a binary mixture with B is proportional to the mole fraction of A only if K_{AA} equals K_{AB}. This remarkable result [98] is shown in Fig. 4-1 for the simple case of a binary gaseous mixture. Analysis of a multicomponent gaseous mixture by measurement of their peak intensities would be very easy if it were not for the broadening effects of the intermolecular collisions. Two simple cases may be discussed. If the component A is present in extremely small concentration, the peak intensity from Eq. (4-2) is given by

$$\alpha_{\nu_{0_A}} = \frac{K'p_A}{K_{AB}}, \qquad (4\text{-}4)$$

where K_{AB} is the dominant parameter. However, when A becomes the chief constituent of the mixture, that is $K_{AA}p_A > K_{AB}p_B$, one may derive the equation

$$\alpha_{\nu_{0_A}} = \frac{K'}{K_{AA}} [1 - \frac{K_{AB}}{K_{AA}} (1 - p_A)], \qquad (4\text{-}5)$$

in which case the peak absorption coefficient is a linear function of the mole fraction of the component A. The peak-intensity measurement can still be a useful analytical technique in certain selective cases where the "K_{AX}" terms are all essentially equal. The line width is then independent of the changes in composition of the mixture, and peak-intensity data can be utilized for accurate comparison of concentrations. Some of these situations are discussed briefly in the paragraphs that follow.

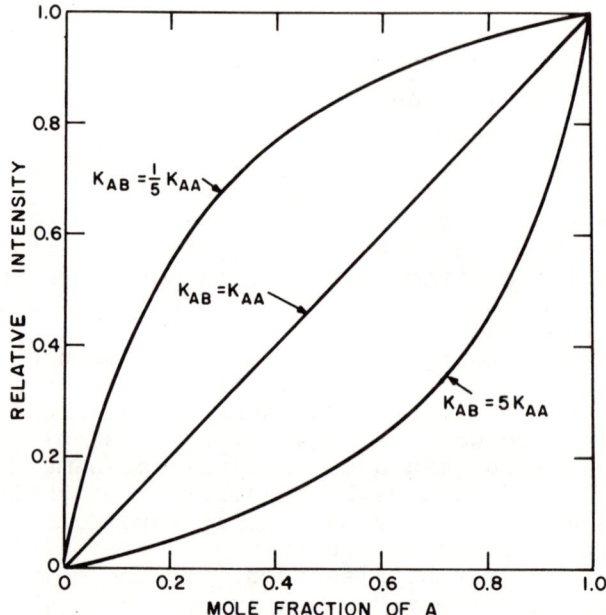

Fig. 4-1. Calculated peak intensities for rotational transition of a gas, A, as function of its mole fraction in binary mixture with component B for different cases of variations in intermolecular line-broadening parameters: K_{AA} = collision-broadening parameter of molecule A for A; K_{AB} = collision-broadening parameter of molecule B for A. (With permission from New York Academy of Sciences.)

$K_{AA} = 5K_{AB}$, $K_{AA} + K_{AB}$, and $K_{AA} + 1/5\ K_{AB}$.

Isotopic substitution or vibrational excitation in general have negligible effects on collision efficiencies of the molecules since the charge distribution of a molecule does not change substantially. Study of peak intensities of transitions may be useful in quantitative determination of the relative proportions of various isotopic forms present in a sample of a molecular species in the vapor phase. If the vibrationally excited states (v = 1, 2, 3, etc.)

PEAK INTENSITY OF ROTATIONAL TRANSITIONS

and the ground state (v = 0) give separate rotational spectra, the peak-intensity method can be used to determine the population in each of the states in a quantitative manner.

If sensitivity for detection of a given component is sufficient, a gaseous mixture may be diluted with a large excess of a nonpolar gas such as argon to minimize the complications due to different line-broadening properties of the molecules of the individual substances. Calibration procedures would compare the measured peak intensity of a transition of a component gas in a binary mixture with argon gas of known composition with the observed peak intensity of the same transition in the gas mixture containing an excess of argon for the determination of mole fraction of the component gas in the mixture.

4-2. INTEGRATED INTENSITY COEFFICIENTS OF ROTATIONAL TRANSITIONS

The integrated absorption coefficient for the rotational transition $J, K_-, K_+ \rightarrow J', K'_-, K'_+$; α_{int}, is given by Eq. (1-34) and can be expressed in a slightly modified form by the equation

$$\alpha_{int} = \pi \Delta \nu \, \alpha_{\nu_0} \, cm^{-1}s^{-1} = \pi U \nu_0^2 \cdot X \cdot p , \qquad (4-6)$$

where

X = mole fraction of the molecular species;

p = total gas pressure; and

$$U = \frac{8\pi^2 F_{J, K_- K_+} \mu_F^2 \lambda_F}{3ck^2 T^2 (2J + 1)} .$$

The term U, whose value is determined by the values of the parameters F_{J, K_-, K_+}, μ_F, λ_F, c, k, J', K_-, K_+, defined in Section 1-5 can be evaluated for the temperature T. It should be mentioned that U is a constant at a particular temperature for a given transition of a molecule in a particular vibrational state. The integrated intensity of a line can be determined by measuring the area under the spectral line. The integrated intensity coefficient, α_{int}, is useful for conducting both relative and absolute intensity measurements for quantitative analysis.

4-3. TECHNIQUES OF MEASUREMENT OF PEAK INTENSITIES AND INTEGRATED INTENSITIES WITH ILLUSTRATIVE APPLICATIONS

Methods now exist that allow an analyst to measure with confidence the relative peak intensities of absorption over a reasonably broad range of frequencies.

4-3-a. Direct Comparison Method

Determination of the $^{37}Cl/^{35}Cl$ isotope ratio by peak-intensity measurements on rotational transitions of $^{37}ClCN$ and $^{35}ClCN$ molecules by Townes, Holden, et al. [99] was one of the early applications of microwave spectroscopy in a quantitative manner. They obtained a value of 0.3 for the isotopic ratio in natural abundance.

The work of Baird and Bird [100] on the relationship between the absorption coefficient of a transition and the crystal response in Stark modulated microwave spectrograph shows the existence of a direct proportionality between the coefficient, α_{v_0}, and the component of crystal current at the modulation frequency, provided that the direct crystal current is maintained constant. To make the lines approximately equal in height for measurement, Esbitt and Wilson [101] conducted peak-intensity determinations by measuring the heights of lines directly at the same modulation voltage by using a calibrated step-attenuator, leading to the detector. In addition, microwave attenuators were used to maintain the direct crystal current at the same level during the measurements. They have determined $^{12}C/^{13}C$ isotope ratios in ^{13}C-enriched samples of methylformate by measuring the peak intensities of the E-lines (internal rotation components) of $1_{01} \rightarrow 2_{02}$ and $1_{11} \rightarrow 2_{12}$ rotational transitions and found a value of 0.635 for $^{12}C/^{13}C$ ratio with a 2.8% standard deviation as compared to a mass-spectrometric determination of 0.644 with 3% uncertainty.

Scharpen and Rasukolb [43] have determined the concentrations of five propene-d_1, C_3H_5D, species present in the samples obtained from catalyzed hydrogen-deuterium exchange reactions of propene by using the method of peak-intensity measurements on a HP Model 8460A Stark modulated microwave spectrometer. Quantitative determination of deuterium distribution in the reaction products provided, for the first time, detailed data that could be utilized to elucidate the mechanism of the chemical reaction. Previous attempts at analyses of propene-d_1 samples by IR and NMR spectroscopy have been inadequate for the purpose. All five subspecies of the monodeutero propene, formed by substitution at one of the numbered hydrogen positions as shown in Fig. 4-2, give rise to separate microwave spectra [102]. A HP 8460A Model spectrometer, interfaced to a HP Model 2116B computer and equipped with facility for automated scan over large frequency regions, was used in Scharpen and Rasukolb's study [43].

By combining Eqs. (1-30), (1-31), and (1-32), the detailed expression for peak absorption coefficient, α_{ν_0}, is given [43] as

$$\alpha_{\nu_0} = N \frac{8(\pi h)^{3/2}}{3c(kT)^{5/2}} (ABC)^{1/2} \cdot F_V \cdot \exp\frac{-(\varepsilon_1 - \varepsilon^0)}{kT} \cdot \frac{(2J+1)\mu_{ij}^2}{\Delta\nu} \quad cm^{-1}, \quad (4-7)$$

where

h = Planck's constant;
N = number of molecules per cm^3;
k = Boltzmann constant;
A,B,C = rotational constants;

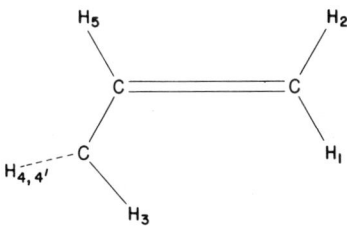

Fig. 4-2. Molecule of propene with numbered hydrogen sites.

F_v = fraction of molecules in occupied vibrational state;

ε_1 = energy of lower rotational state, J, K_-, K_+ of transition;

ε^0 = energy of lowest energy state available to molecule; and

μ_{ij} = dipole-moment matrix element for transition.

The output signal, S, can also be expressed in volts by the equation

$$S = G\alpha_{\nu_0} \ell \qquad (4-8)$$

where

G = a constant if rectified d.c. current from microwave detector is fixed at a certain value; and

ℓ = effective path length in centimeters for absorption, related to physical length of wave guide, L, by the relation $\ell = L(\lambda_g/\lambda)$, where λ_g and λ are the microwave wavelengths in waveguide and free space, respectively.

Equations (1-33), (4-2), and (4-8) can be combined to yield the result

$$p = KS_p \frac{\Delta\nu(\lambda/\lambda_g)}{T^{3/2}} \qquad (4-9)$$

where

K = a proportionality constant;

S_p = peak absorption signal; and

p = partial pressure of component in mixture.

Sample pressures of 13 Pa, detector crystal current of 100 µA, and temperature of 298 K were chosen as operating conditions. The ratios of signals, S_p, from each of the five propene-d_1 species to that from the propene-d_0 species in the samples were experimentally determined. The ratio of peak signal intensity of the propene-d_0 line in the partially deuterated sample to that in pure propene was experimentally determined. This ratio gave directly the fractional amount of C_3H_6 in each sample. Scharpen converts the experimentally determined peak signal ratios into the actual partial pressure or concentration ratios by multiplying these data by factors that take into account the differences in quantities ℓ, $(ABC)^{1/2}$, F_v, μ_{ij}^2, ν^2, and by multipliers determined experimentally in an empirical manner to correct for the effect of power saturation on (HWHH), $\Delta\nu$, and for hindered internal rotation effects. The correction factors can be computed or determined experimentally with a high degree of precision.

The observed percent distribution of propene-d_1 species in samples obtained in the manner outlined in the preceding paragraphs is given in Table 4-1. A comparison of results obtained by microwave and mass-spectrometric methods is given in Table 4-2.

The isomerization of n-butene over deuterated p-toluene sulfonic acid was investigated by Sakurai, Kaneda, et al. [103] by microwave rotational spectroscopy. The isomers 1-butene and 2-butene in the reaction product were separated by fractionation at liquid-N_2 temperatures. The samples thus obtained were analyzed by microwave spectroscopic technique. Table 4-3 gives the spectral frequencies of the transitions of the monodeutero 1-butene and monodeutero(cis)2-butene (structures shown in Fig. 4-3) and the computed correction coefficients for obtaining true ratio of partial pressures from the measured peak-intensity ratios by a simple arithmatic operation.

Another quantitative analysis of products of catalyzed deuteration reaction of propene in the presence of D_2SO_4 and D_2PO_4 has been conducted by Kondo, Saito, et al. [104]. The techniques followed were similar to those employed by Scharpen and Rasukolb [43] described earlier for propene-d_1 species. To monitor the kinetics of deuteration reactions of propene in the presence of D_2SO_4 and D_2PO_4, the distribution of geometrical isomers of propene-d_1 and -d_2 were determined during the course of the chemical reaction by using the peak-intensity method. The number of hydrogen-atom positions in propene is shown in Fig. 4-2. The numbering of D atoms would denote the place of the original hydrogen atom in the molecule. The quantitative analyses of the geometrical isomers of propene-d_1 and propene-d_2 were conducted by measuring

TABLE 4-1. OBSERVED PERCENT DISTRIBUTION[a] OF PROPENE-D$_1$ SPECIES IN CATALYZED DEUTERIUM-EXCHANGE SAMPLES (Reproduced with permission of American Chemical Society.)

Species	Sample Identification				
	Pt(1)	Pt(2)	Rh(1)	Rh(2)	Ni(1)
CH$_3$CDCH$_2$	58.0(0.7)[b]	57.8(0.6)	30.7(0.4)	25.4(0.3)	16.8(0.4)
CH$_2$DCHCH$_2$[c]	26.2(0.5)	26.6(0.3)	34.2(0.4)	40. (0.5)	53.1(0.7)
cis-CH$_3$CHCHD	7.9(0.2)	8.0(0.2)	17.5(0.3)	17.2(0.3)	15.3(0.3)
trans-CH$_3$CHCHD	8.0(0.2)	7.7(0.1)	17.6(0.3)	17.0(0.3)	14.8(0.3)

[a]Transitions with absorption coefficients as low as 5×10^{-9} cm^{-1} were measured in this study. The relative standard deviation of the analytical result was about 2%.

[b]The quantities in the parentheses are the standard deviation.

[c]Data here are the sum of values determined for the sym- and asym-CH$_2$DCHCH$_2$ species.

TABLE 4-2. PERCENTAGES OF PROPENE-D_0 AND PROPENE-D_1 FROM MICROWAVE AND MASS-SPECTROSCOPIC DATA (Reproduced with permission of American Chemical Society.)

Catalyst	Propene-d_0		Propene-d_1	
	Microwave	Mass Spec.	Microwave	Mass Spec.
Pt(1)	73.7(0.7)[a]	75	22.6(0.3)	23
Pt(2)	63.2(0.7)	64	29.3(0.4)	30
Rh(1)	52.7(0.7)	55	31.9(0.5)	32
Rh(2)	21.3(1.0)	27	31.6(1.5)	31
Ni	90.8(0.8)	89	10.4(0.1)	10

[a]The quantity in parentheses is the standard deviation. Column 1 indicates the run number of the experiment.

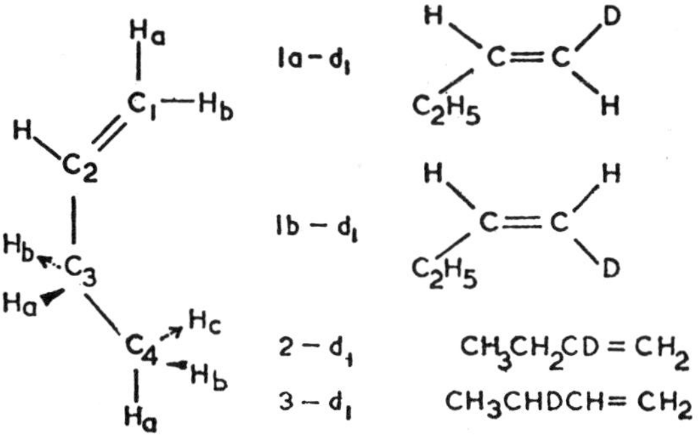

Fig. 4-3. Structural formula of monodeutero 1-butenes.

TABLE 4-3. ROTATIONAL SPECTRA OF SUBSTITUTED MONO-DEUTERO SPECIES OF 1-BUTENE (SKEW FORM) AND (CIS) 2-BUTENE AND COMPUTED CORRECTION COEFFICIENTS (Reproduced with permission of the American Chemical Society.)

Species	Transition	Frequency/ (MHz)	Correction Coefficient
1-Butene and its Monodeuterated Species (Skew Form)			
1a-d_1	3_{12}-2_{11}	23,392.20	1.172
1b-d_1	3_{12}-2_{11}	23,912.96	1.123
1-d_1	3_{12}-2_{11}	24,493.02	1.062
3a-d_1	3_{12}-2_{11}	24,479.56	1.068
3b-d_1	3_{12}-2_{11}	24,633.84	1.044
4a-d_1	3_{12}-2_{11}	23,419.28	1.161
4b-d_1	3_{12}-2_{11}	23,959.08	1.115
4c-d_1	3_{12}-2_{11}	24,134.66	1.104
d_0	3_{13}-2_{11}	24,487.28	1.000
cis-2-Butene and its Monodeuterated Species			
d_0	6_{24}-6_{15}	28,666.54	1.000
1a-d_1	8_{26}-8_{17}	28,398.54	1.237
1b-d_1	8_{26}-8_{17}	28,751.68	1.610
2-d_1	8_{26}-8_{17}	28,270.52	0.832

the peak intensities of $0_{00} \rightarrow 1_{01}$ rotational transitions, which are given in Table 4-4. A conventional 110-kHz sinusoidal Stark modulated microwave spectrometer, equipped with ferrite isolators at both ends of the 3-m-long Stark cell, was used. Measurements were made at dry-ice temperatures. The spectrometer sensitivity was about 1×10^{-10} cm^{-1}.

The absorption coefficient, α_{ν_0}, is given by Eq. (4-7). The measured peak-intensity ratios were converted to ratios of partial pressures by dividing with the total correction coefficients computed from

TABLE 4-4. ROTATIONAL TRANSITIONS OF PROPENE-D_1 AND PROPENE-D_2 SPECIES (Reproduced with permission of American Chemical Society.)

Species	Frequency, MHz	Species	Frequency, MHz
D_6	17,139.00	D_4D_6	16,104.01
D_3	16,832.95	D_3D_4	15,864.90
D_1	16,769.76	D_2D_6	15,835.80
D_4	16,377.12	D_1D_4	15,751.82
D_2	16,090.18	D_1D_2	15,550.79
D_3D_6	16,542.01	D_2D_3	15,534.33
D_1D_6	16,489.77	D_4D_5	15,482.30
D_1D_3	16,218.95	D_2D_4	15,140.28

factors such as the F_V, $(ABC)^{1/2}$, μ_a^2, ν_0^2, and so on. The parmeters such as F_V, the μ_a-dipole moment component, $(ABC)^{1/2}$, and so forth have slightly different values for the isotopic species, and hence their variations have to be taken into account. For example, the main contribution to F_V (the fraction of molecules in the lowest vibrational state) was found to be from the torsional vibration mode of 193.6 cm^{-1} for C_3H_6. Using the value of kT = 142 cm^{-1} one would calculate $F_V(CH_3)/F_V(CH_2D) = 1.0740$. The half line widths at half maximum intensity (HWHH) of the lines of transition of d_1- and d_2-species were considered constant. The correction coefficient for propene-d_1 and propene-d_2 species are given in Table 4-5.

4.3.b. Use of PIN-diode Signal Calibrators

The PIN diode [41] in the HP Model 8460A spectrometer provides an easy approach to the measurement of peak intensities. The variable square wave power modulated at 33.33-kHz frequency (same as the Stark modulation frequency) from the PIN diode can be added to (in phase) or subtracted from (out of phase) the absorption signal in the Stark cell with the help of the phase shifter. The power provided by the PIN diode is calibrated by using standard attenuators.

TABLE 4-5. CORRECTION COEFFICIENTS FOR THE INTENSITY OF RATIOS BETWEEN PROPENE-D_1 ISOMERS AND BETWEEN PROPENE-D_2 ISOMERS (Reproduced with permission of American Chemical Society.)

Ratios	F_v	$(ABC)^{1/2}$	μ_a^2	ν_0^2	Total Corrections
$D_2:D_4$	1.0740	1.0117	0.9961	0.9653	1.0448
$D_1:D_3$	1.0740	0.9922	1.0196	0.9925	1.0783
$D_6:D_3$	1.0740	0.9871	1.0090	1.0367	1.1089
$D_1D_4:D_3D_4$	1.0625	0.9926	1.0196	0.9858	1.0600
$D_1D_2:D_2D_3$	1.0740	0.9942	1.0197	1.0021	1.0911
$D_2D_3:D_4D_5$	1.0625	1.0098	0.9851	1.0067	1.0640
$D_4D_6:D_3D_4$	1.0625	0.9871	1.0090	1.0304	1.0904
$D_2D_6:D_3D_4$	1.1411	0.9939	1.0050	0.9963	1.1355
$D_1D_6:D_3D_6$	1.0740	0.9918	1.0197	0.9937	1.0793

A favorable situation would require the PIN-diode power to be of phase opposite to that of the absorption signal. The PIN-diode calibrated signal can be varied over wide ranges of microwave frequencies. This provides an unusual opportunity for the comparison of peak intensities of transitions spread over a wide range of frequencies. This method also provides for an accurate and reproducible calibration of instrument performance with frequency variation.

4.4. TECHNIQUES OF RELATIVE AND ABSOLUTE INTENSITY MEASUREMENTS

Peak intensity data without the HWHH data are inadequate for relative or absolute intensity measurements because of line-broadening effects with the sample gas at pressures of about 1 to 15 Pa.
The determination of both peak intensity and HWHH $\Delta\nu$, or the integrated intensities constitutes valid techniques for quantitative analysis. Crable [105]

has experimentally studied the variation of line height, width by height, and area under a spectral line for gases such as SO_2, CH_2O, acetone, and OCS in pure state as a function of sample pressure. The observed variation of peak intensities, height by width, and the area under a spectral line with pressure of the sample gas are shown in Figs. 4-4 through 4-7. The data clearly establish that the product of peak intensity and line width or the area under a spectral line is a linear function of sample pressure. However, line height was found to vary linearly with pressure only in low-pressure regions (0.1 to 1 Pa).

An appropriate equation can be derived from Eq. (4-6), which would directly give the ratio of partial pressures from intensity parameters. The experimentally observed intensities of spectral lines from any

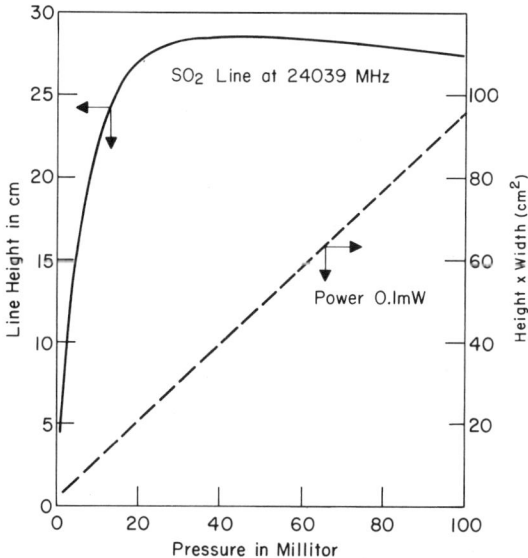

Fig. 4-4. Observed variation of peak intensity (line height) as well as the product of peak intensity and line width, with pressure of SO_2 gas; SO_2 transition at 24,039.6 MHz was used for this investigation. (With permission of author, G. F. Crable [106].)

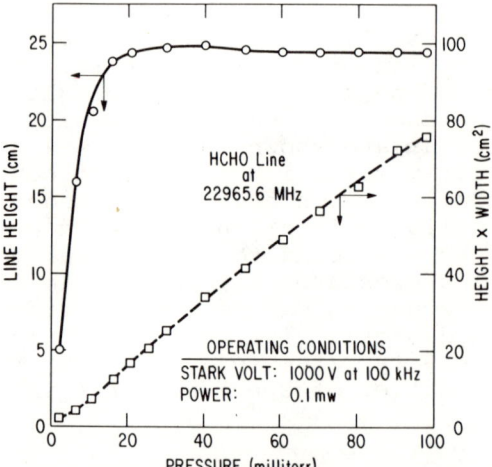

Fig. 4-5. Observed variation of peak intensity and line width, with pressure of HCHO gas for transition at 22,965.6 MHz. (After Crable [106].)

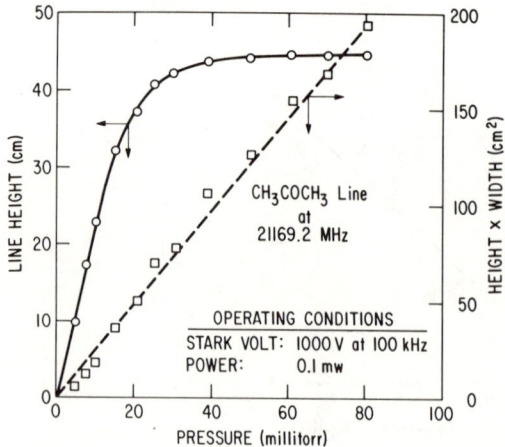

Fig. 4-6. Observed variation of peak intensity (line height) as well as product of peak intensity and line width with pressure of acetone gas for transition at 21,169.2 MHz. (After Crable [107].)

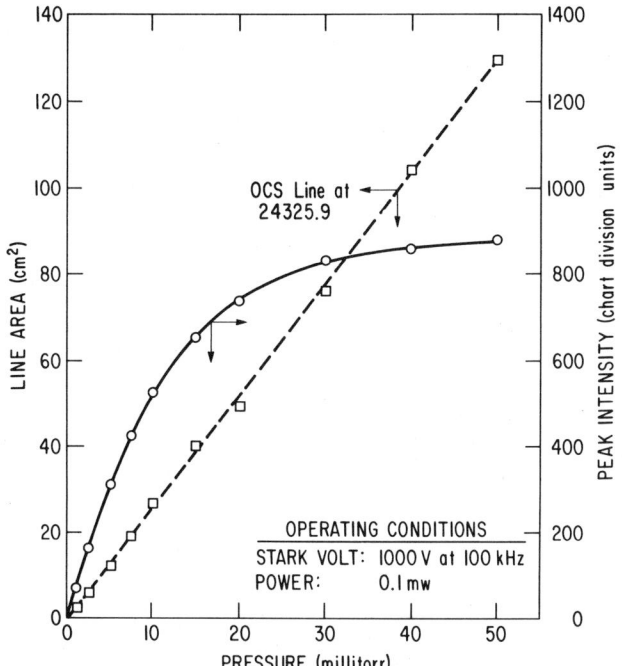

Fig. 4-7. Observed variation of peak intensity (line height) and area under line with pressure of OCS gas for transition at 24,325,9 MHz. (After Crable [107].)

two components of a gaseous mixture can be utilized to directly give the ratio of partial pressures of the gases according to the equation

$$\frac{(\alpha_{\nu_0} \cdot \Delta\nu)_1}{(\alpha_{\nu_0} \cdot \Delta\nu)_2} = \frac{(\alpha_{int})_1}{(\alpha_{int})_2} = \frac{U_1 \cdot \nu_{01} \cdot p_1}{U_2 \cdot \nu_{02}^2 \cdot p_2} \qquad (4-10)$$

where

$\alpha_{\nu_{01}}, \alpha_{\nu_{02}}$ = peak absorption coefficients of transitions from components 1 and 2;

$(\alpha_{int})_1, (\alpha_{int})_2$ = integrated absorption coefficients of transitions from components 1 and 2; and

$\nu_{01}\ \nu_{02}$ = peak frequencies of observed transitions from components 1 and 2.

The terms U_1 and U_2 can be calculated from known or computed values of the parameters, c, k, T, $F_{J,K-,K+}$, μ_F, λ_F, J, and so on. The parameters themselves have been defined in Section 1-5.

Relative intensity measurements on a number of gaseous components of a gaseous mixture can be conducted by: (1) a determination of peak intensities and HWHH or (2) a measurement of integrated absorption coefficient. Various experimental procedures are available for conducting both relative and absolute intensity measurements. The most commonly used or promising techniques are most easily accomplished with automated commercial spectrometers. Some of the techniques are discussed in the following paragraphs.

4.4.a. Bridge Method

The HP Model 8460A microwave spectrometer is provided with a bridge arrangement [106] as shown in Fig. 2-4. The arrangement allows the microwave power level in the absorption cell to be altered without affecting total power level arriving at the crystal detector. The phase angle between the sample arm and bridge arm is adjusted so that the microwave electric fields add at the junction of two arms before the detector. Power saturation can be avoided or at least known through this arrangement. The following procedure may be used to determine the extent of microwave power absorbed by the molecules at the peak frequency of the transition.

The absorption cell is pumped out and the spectrometer tuned to the peak absorption frequency ν_0 of a gas. The absorption cell attenuator setting, A_i (decibels), with no power going through the bridge, is read for a constant crystal current. The sample is then introduced at a certain chosen pressure. A low level of power through the cell is selected to avoid power saturation, and the absorption cell attenuator setting, A_b (in decibels), is noted. The bridge attenuator is set to give the same total crystal current, i_0, at the detector as before. The signal, S_1, is now given by the equation

$$S_1 = G \cdot \alpha_{\nu_{01}} \cdot L \cdot 10^{(A_i - A_b)/20} \qquad (4\text{-}11)$$

where the term G represents the overall spectrometer gain and L is the length of the absorption cell. The HWHH is determined by following a slightly different procedure. The sample pressure and the Stark cell attenuator settings are not changed, and the recorder pen is allowed to make a trace on the strip chart with the Stark field on. After making this trace, the Stark field is turned off. The pen comes to rest on the baseline, which is now traced by the pen; therefore, the midline between the traces is the half-intensity line. The Stark field is turned on and the pen is brought to rest on the top trace. The frequency is now scanned to the high side until the pen crosses the half-intensity line. The intersection of the line shape with the half-intensity line now allows measurement of the frequency separation to give the HWHH, $\Delta\nu_1$. A similar procedure may be followed to determine the microwave signal, S_2, and the corresponding HWHH, $\Delta\nu_2$, for a chosen transition from the second molecular species in the mixture. The ratio

$$\frac{S_1 \cdot \Delta\nu_1}{S_2 \cdot \Delta\nu_2}$$

equals the ratio of partial pressures p_1/p_2.

4.4.b. Peak Height and HWHH, $\Delta\nu$, from Spectral Traces on Strip Chart for Determination of Relative Intensities

Measurement of peak height and HWHH, $\Delta\nu$, from a spectral trace on a strip chart is a straightforward procedure once a satisfactory baseline is established. Proper control of experimental conditions and spectrometer functions is, however, extremely important to guarantee reproducible and accurate determinations of relative intensities. Important factors are to: (1) avoid power saturation, (2) maintain a steady flat baseline by avoiding pickup, (3) apply a constant Stark voltage during measurements, (4) assure complete modulation, and (5) avoid spectral overlaps.

The ratio of the products, $\alpha_{\nu_0} \cdot \Delta\nu$, for various transitions from the components of a gaseous mixture would directly yield the ratio of partial pressures according to Eq. (4-10).

4.4.c. Measurement of Integrated Intensities

The integrated intensity, as mentioned earlier, is given by the area under the spectral line trace and is simply related to partial pressures by Eq. (4-6).

The measurement of areas under the spectral lines can be conducted such as in the method used in the commercial CSI spectrometer [44] equipped with an automatic electronic integration system. Some of the sources of errors in the determination of areas representative of the actual integration coefficients are: (1) inability to define a zero baseline for the Lorentzian line shape, (2) overlap by neighboring transitions or their Stark lobes, and (3) incomplete modulation. Distortion of line shape due to incomplete separation of Stark lobes from the line appears to present a serious problem. However, with the instrument of British design [44] this problem is readily solved.

The automatic electronic integration system provided with the instrument can measure the area under portions of line contour that extend from the peak point to a point lower down that is a known fraction of the peak height above the baseline (i.e., 90.5%, 79.4%, 65.5%, and 50% with the CSI spectrometer). By using the on-line computer and assuming a Lorentzian line shape, the net normalized area under the spectral line is computed for the different choices of end points of integration and printed out. The integration end point is chosen appropriately so that integrated absorption coefficient obtained is relatively free from distortion due to overlapping Stark lobes or lines.

4.4.d. Absolute Intensity Measurements

Measurement of peak intensity, α_{ν_0}, and HWHH or the measurement of the area under an absorption line for transitions of the sample under standardized conditions are essential prerequisites for determining the absolute intensities.

Almost all measurements reported in the literature have been based on some sort of calibration procedure.

TECHNIQUES OF RELATIVE AND ABSOLUTE INTENSITY

For example, the concentrations may be determined by measuring α_{ν_0}, $\Delta\nu$, and p for a given transition in a standard sample. The same quantities are then measured under the same instrumental as well as experimental conditions in the gas mixture.

Study of quantitative determinations of thermodynamic functions [106] and chemical equilibria [107] by absolute and relative intensities measurements is reviewed as follows. The works emphasize the powerful potential of microwave spectroscopy as an analytical tool. The standard enthalpy of reaction at absolute zero may be obtained directly if the intensity of a rotational transition of each species involved in a chemical reaction is measured, provided that the data on the rotational spectrum and dipole moment of the molecular species are available. This technique was used by Saegebarth and Wilson [108] to determine the energy difference between two rotamers of fluoroacetyl fluoride, CH_2FCFO. This approach was also used by Varma and Curl [107] to determine the energy difference between the two rotamers of HNO_2 and DNO_2 as well as to determine the energy difference between the ground states of products and ground states of reactants for the reaction

$$N_2O_3 + H_2O \rightleftarrows 2HNO_2 .$$

The product of unsaturated peak absorption coefficient, α_{ν_0}, and HWHM, $\Delta\nu$, is given by a slightly modified form of Eq. (1-33):

$$\alpha_{\nu_0} \cdot \Delta\nu = \frac{8\pi^2 \nu_0^2 |<J"_{K_-",K_+"}, F"|\mu|J'_{K_-',K_+'}, F'>|^2}{3v_g kTQ_{int}Q_{ns}}$$

$$\omega_I N \exp \frac{-E"}{kT} \quad cm^{-1} \; s^{-1} \qquad (4-12)$$

where

v_g = group velocity of light given by expression $(c \cdot \lambda/\lambda_g)$ (where c = velocity of light, λ and λ_g are wave-lengths of microwave radiation in free space and in absorption cell);

ω_I = nuclear spin weight for any unresolved nuclear hyperfine levels;

Q_{ns} = nuclear spin partition function of unresolved nuclear spins;

N = number of molecules per cm^3 of species involved;

E" = energy of lower level involved in transition;

k = Boltzmann constant;

T = temperature in degrees Kelvin; and

$<J''_{K''_-,K''_+},F''|\mu|J'_{K'_-,K'_+},F'>$ = reduced dipole moment matrix element of the particular quadrupole hyperfine transition observed for $J'_{K'_-,K'_+} \leftarrow J''_{K''_-,K''_+}$ rotational transition.

The energy difference between the two lower levels of the transitions from cis and trans HNO$_2$ and DNO$_2$ species can be expressed in terms of intensity parameters by the equation

$$E_c(J''_c) - E_t(J''_t)$$

$$= kT \frac{(\alpha_{\nu_0} \cdot \Delta\nu)_t |<J''_{K''_-,K''_+},F''|\mu|J'_{K'_-,K'_+},F'>_c|^2 \nu_g^t \nu_{0c}^2}{(\alpha_{\nu_0} \cdot \Delta\nu)_c |<J''_{K''_-,K''_+},F''|\mu|J'_{K'_-,K'_+},F'>_t|^2 \nu_g^c \nu_{0t}^2} \quad (4-13)$$

where the labels c and t refer to cis and trans forms.

The intensity of a transition was measured by recording the line and taking the product of peak height and full width at half-height. The effect of power saturation was minimized by working at a relatively high pressure of ∼70 Pa. The line widths of all observed lines were approximately equal, assuring similar relaxation times. The measurements were made using a HP Model 8460A spectrometer in the frequency region 26.5 to 40 GHz. The absorption cell was kept at an average temperature of -49°C for determination of relative intensities of reactants and products for the reactions

$$N_2O_3 + D_2O \rightleftarrows 2DNO_2, \qquad (4-14)$$

$$N_2O_3 + HDO \rightleftarrows HNO_2 + DNO_2 . \qquad (4-15)$$

The absorption cell was kept at about 30°C for intensity measurements on cis and trans HNO_2 and DNO_2 species.

A few typical measured intensity parameters and results on cis-trans forms of HNO_2 and DNO_2 are given in Table 4-6a. The calculated values of the square of reduced matrix elements in units of D^2 and the energy of the lower rotational state of the transitions of the molecules are given in Table 4-6b. The presence of hyperfine structure arising from the nuclear quadrupole coupling due to ^{14}N causes certain complications. Transitions for measurement were carefully chosen for this reason. Only well-resolved single hyperfine components or lines that had unresolved hyperfine structure (all components within 0.3 MHz) were chosen for measurements. The lines were fully modulated and free from overlap.

TABLE 4-6a. INTENSITY MEASUREMENTS AND CIS-TRANS ENERGY DIFFERENCES (Reproduced with permission of American Chemical Society)

Run Number	Temp, °C	cis line[a]	trans Line[a]	Peak Height, mm	$\Delta\nu$, MHz	$E_c - E_t$ cal/mol
HNO_2 (1)	30.1	2		79.0	0.580	
			3	75.0	0.630	534
			4	36.5	0.550	436
HNO_2 (2)	30.1	1		21.5	0.460	
			3	75.0	0.630	453
			4	36.5	0.550	354
DNO_2 (1)		5		121	0.394	
			7	141	0.53	499
DNO_2 (2)		5		93	0.498	
			6	77	0.515	521

[a]Please refer to Table 4-6B for line designation.

TABLE 4-6b. TRANSITIONS CHOSEN FOR INTENSITY MEASUREMENTS AND CALCULATED VALUES OF REDUCED MATRIX ELEMENTS AND ENERGY OF LOWER LEVEL OF TRANSITIONS (Reproduced with permission of American Chemical Society.)

Species	Line No.	Rotational transition $J'K'_-,K'_+ \leftarrow J''K''_-,K''_+$	Hyperfine $F' \leftarrow F''$	Frequency ν_0, MHz	$\|\langle J''_{K_-,K_+,F''}\|\mu\| J'_{K_-,K_+,F'}\rangle\|^2$ D^2	$E_{J''_{K_-,K_+}}$ THz
cis-HNO_2	1	$10_{1,10} \leftarrow 9_{2,7}$	$10 \leftarrow 9$	32,739.59	2.00	1.368
	2	$18_{4,14} \leftarrow 19_{3,17}$		35,529.60	16.92	5.322
trans-HNO_2	3	$11_{2,9} \leftarrow 12_{1,12}$	$11 \leftarrow 12$	36,503.80	1.95	1.854
	4	$11_{2,9} \leftarrow 12_{1,12}$	$\{\begin{array}{l}10 \leftarrow 11\\12 \leftarrow 13\end{array}$	36,501.11	3.92	1.854
cis-DNO_2	5	$15_{4,11} \leftarrow 16_{3,4}$		30,083.5	12.51	3.777
trans-DNO_2	6	$11_{2,9} \leftarrow 12_{1,12}$	$11 \leftarrow 12$	38,976.87	1.93	1.735
	7	$11_{2,9} \leftarrow 12_{1,2}$	$\{\begin{array}{l}10 \leftarrow 11\\12 \leftarrow 13\end{array}$	38,974.32	3.87	1.735

The equilibria for the related reactions (4-14) and (4-15) were investigated to determine the standard enthalpy ΔH_0° for the reaction,

$$N_2O_3 + H_2O \rightleftarrows 2HNO_2 . \qquad (4-16)$$

The standard enthalpy ΔH_0° for any of the reactions is given by the equation

$$\Delta H_0^\circ = -\Delta E_J - RT \ln \frac{R_V R_I}{R_\nu^2 R_\mu^2 R_\omega} + \frac{3}{2} RT \ln R_M \qquad (4-17)$$

where for reaction (4-14), for example,

$$\Delta E_J = 2(E_0 - E^{(0)})_{DNO_2} - (E_0 - E^{(0)})_{N_2O_3}$$
$$- (E_0 - E^{(0)})_{D_2O} \qquad (4-18)$$

and $(E_0 - E^{(0)})$ is the energy of the lower level involved in the transition relative to the lowest energy of the molecule. The R values are the ratios of same structure as the equilibrium constant for the corresponding reaction. For example, R_V for the reaction (4-14) is given by the relation

$$R_V = \frac{(v_g^{DNO_2})^2}{v_g^{N_2O_3} v_g^{D_2O}} \qquad (4-19)$$

where the v_g values are the group velocities of radiation in the cell at the frequencies of the corresponding rotational transition.

The ratio of intensities, R_I, was determined from relative intensity measurements on the gaseous reactants and products for the reactions described by reactions (4-14) and (4-15). The standard enthalpy, ΔH_0°, for reactions were then obtained by substituting the values of the ratios R_I, R_V, R_ν, R_μ, R_ω, and R_M in appropriate equations.

A simple procedure, based on the measurement of: (1) peak intensity, α_{ν_0}, by the "bridge method" and (2) the HWHH, $\Delta \nu$, by observing the frequency displacement, $\delta \nu (\delta \Delta = \Delta \nu)$, of the penrecorder on the strip chart from the peak during the line scan (to either high- or low-frequency side) required to reduce the maximum intensity by a factor of 2 at constant power, developed by Curl, Ikeda, et al. [106] was used for determining absolute intensities of transitions. An additional feature of the absolute intensity procedure involves the calibration of the spectrometer gain, GL, in Eq. (4-11) by intensity measurement of a reference substance.

A summary of the work of Curl, Ikeda, et al. [106] on the determination of thermodynamic functions of pure substances is given as follows to illustrate the usefulness of absolute intensity determination as a method for quantitative chemical analysis.

Equation (4-12) can be applied to a rotation transition, $J'_{K'_-,K'_+} \leftarrow J"_{K"_-,K"_+}$, unaffected by splittings due to nuclear quadrupole or internal rotation effects. The reduced dipole-moment matrix element of Eq. (4-12) is now replaced by the simple dipole-moment matrix element $|<J"_{K"_-,K"_+}|\mu_F|J'_{K'_-,K'_+}>|$, where μ_F stands for the a,b,c components of the dipole moment ($\mu_F = \mu_a$, μ_b, or μ_c). Experimental determination of the microwave signal, $S[S \propto \alpha_{\nu_0};$ see Eq. (4-12)], as well as the HWHH, $\Delta\nu$, was conducted according to methods outlined before. The quantities ν_o, $|<J"_{K"_-,K"_+}|\mu_F|J'_{K'_-,K'_+}>|^2$, ω_I (nuclear spin weight), and E" can be computed from the assigned rotational spectrum and the dipole moment of the molecule. The experimental value of signal, S, or the quantity $\alpha_{\nu_0} \cdot \Delta\nu$ substituted into the appropriate form of Eq. (4-12) would yield the value of the rotation-vibration partition function, Q_{int}, if the pressure of the pure substance in the cell is known. The free-energy function, $-(G°-H°_0)/T$, in entropy units may be evaluated from the equation

$$\frac{-(G°-H°_0)}{T} = R(\ln Q_{int} + \frac{5}{2} \ln t + \frac{3}{2} \ln M) - 7.283$$

(4-20)

A calibration procedure was followed in actual practice. The parameters S and $\Delta\nu$ were determined for the transitions, as well as the pressure of SO_2. The spectrometer gain, GL, was now evaluated by Eq. (4-11) using values of $\alpha_{\nu_0} \cdot \Delta\nu$ calculated for SO_2 transitions from Eq. (4-12).

Using the gain factor, GL, thus evluated from SO_2 measurements, the values of Q_{int} were now determined for a number of pure substances by using Eqs. (4-11) and (4-12). The results of a typical determination of the free-energy function are given in Table 4-7.

TABLE 4-7. FREE-ENERGY FUNCTION OF CH_2F_2 (Reproduced with permission of American Chemical Society.)

Transition	ν_{obs}, MHz	$-(G°-H_0°)/T$, cal/(mole deg)
$1_{01} \to 1_{10}$	39,892.53	50.30 ± 0.02
$4_{13} \to 3_{22}$	31,543.82	50.37 ± 0.02
$5_{15} \to 4_{22}$	29,268.57	50.35 ± 0.05
$9_{27} \to 10_{1,10}$	30,679.10	50.49
$8_{26} \to 7_{35}$	31,763.69	50.39
$11_{38} \to 12_{2,11}$	32,507.71	50.50
$12_{3,10} \to 11_{47}$	34,755.89	50.36
$6_{25} \to 7_{16}$	39,501.07	50.46
Average		50.40 ± 0.08 at 298.5 ± 0.3 K
Observed		50.39 ± 0.12 ⎫
JANAF Value (interpolated)		50.37 ⎬ at 298.15 K

For analytical methods based on power saturation, measurements of the peak-intensity coefficient, α_{ν_0}, and half line width at half maximum intensity, $\Delta\nu$, as well as the integrated absorption coefficient under conditions that avoid power saturation have been used in a large number of applications. However, such measurements are difficult for extremely weak transitions. The experimental procedures under conditions of power saturation suggested by Harrington [38,39] have distinct advantages in this regard.

The theoretical treatment of Harrington expresses a newly defined Γ-coefficient as a product of two functions, $\Gamma = \eta \cdot \phi$, where η is related to the concentration and ϕ is a power-related parameter. The value of ϕ may be changed by varying the incident power, P_0 [$\phi = f(KP_0)$, where K = saturation coefficient and P_0 is the incident power]. The techniques of chemical analysis proposed by Harrington involve increasing the power until the maximum signal is obtained. This also maximizes ϕ, which is a constant

Measurements of the peak-intensity coefficient, α_{ν_0}

for a given insertion loss. The relation $\Gamma_{max}:\eta:S_{max}:xp$, where p is pressure of the sample gas, x is the mole fraction of the component in a gaseous mixture, and S_{max} is the maximized signal, is valid.

4-5. Γ-INTENSITY COEFFICIENTS FOR CHEMICAL ANALYSIS UNDER CONDITIONS OF POWER SATURATION

The use of Γ-absorption coefficient offers a number of advantages. A maximized signal is measured, enabling the monitoring of concentrations of molecular species that give rise to weak microwave transitions. The maximized spectrometer signal, S_{max}, proportional to the Γ-coefficient, is measured in a simple way since determination of only the peak height or peak intensity is required. The Γ-coefficient is proportional to concentration of the molecular species but is independent of the linewidth. Examination of the properties of the Γ-coefficient as given in the relation

$$\Gamma:S_{max}:xp \tag{4-21}$$

suggests a linear relationship between the maximized signal, S_{max}, and the pressure, p. Furthermore, the value of Γ-coefficient or S_{max} is not affected due to the presence of impurities in the sample, as it is independent of $\Delta\nu$. The two properties are illustrated in Fig. 4-8. The signal, S_{max}, of ethylene oxide is linear with pressure and does not change even when impurities such as water vapor, p-chlorotoluene, or air are added to it.

An obvious disadvantage of the Γ-coefficient is the requirement of high microwave power density in the sample cell. For sample pressures of approximately 5 to 10 Pa and line widths of a few MHz, the power needed for saturation may exceed the output of about 10 mW generally available from the BWOs.

Most of the work done to date on intensity measurements under conditions of power saturation have primarily been devoted to the development of techniques. A number of analytical techniques based on measurement of Γ-absorption coefficients are discussed in the following paragraphs. Results of a few investigations, based on the applications of the technique, are used as illustrative examples. Other examples of this method are given in Chapter 5.

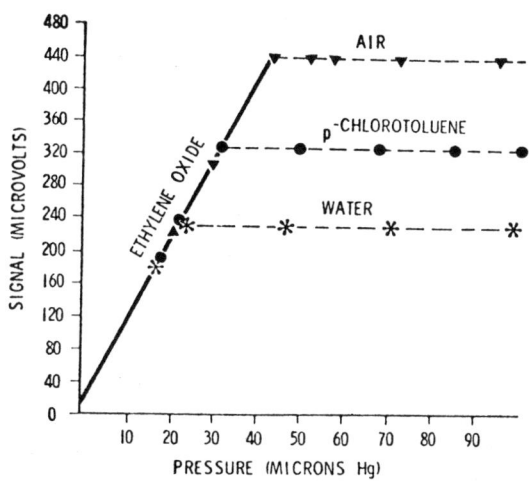

Fig. 4-8. Variations of maximum signal, S_{max}, of ethylene oxide with pressure in presence of impurities (from Harrington [39]).

4.5.a Use of Calibration Procedures for Measurement of Γ-Coefficient

The approximate relation, $\Gamma : S_{max} : xp$, suggests a linear relation between the maximized signal, S_{max}, of a transition from a substance and the pressure of the sample. The linear plot does not pass through the origin, apparently because of the effect of wall collisions at low sample pressures. This necessitates a calibration procedure.

The signal is maximized by increasing the microwave power level suitably for any sample pressure. A calibration curve of S_{max} versus p for a pure gas ($x = 1$) is experimentally obtained. The signal, S_{max}, of the component gas in the mixture is now measured and the pressure directly read from the calibration curve. The mole fraction of the component gas in the mixture is calculated from the measured total pressure of the mixture.

Crable [105] has measured the maximized signal, S_{max}, of the 24325.92-MHz OCS line as a function of pressure and power required to saturate the line. His results are shown in Fig. 4-9. The OCS line was chosen because its high intensity and its narrow width

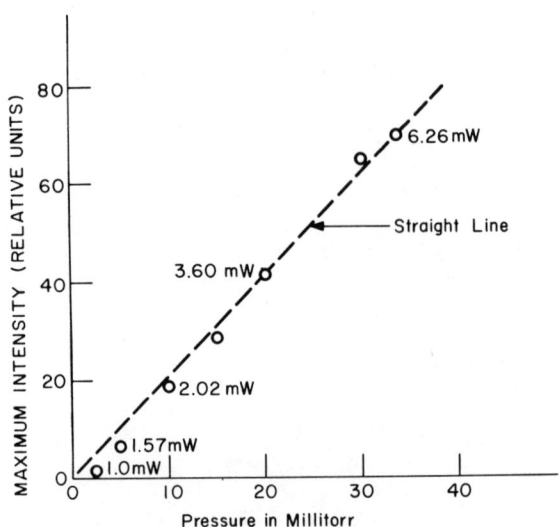

Fig. 4-9. Maximized signal, S_{max}, of OCS transition at 24,325,92 MHz plotted against pressure of gas; powers required to saturate line at different pressures are given.

makes it easy to power-saturate. The signal is seen to vary linearly with the pressure above 1.3 Pa. This provides a very reliable basis for conducting chemical analysis of OCS gas in pure state or in a mixture with nonpolar diluent gas such as nitrogen. A calibration curve showing variation of signal, S_{max}, with the pressure of pure OCS is compared with the intensity obtained, for example, for a mixture with air to determine the percent OCS in the air-OCS mixture.

4.5.b. Use of PIN Diode and Calibration Arm for Measurement of Γ_{max} Intensity

The PIN diode, provided with the HP Model 8460A microwave spectrometer [41], can be used to measure a maximized signal, S_{max}, produced by adjusting the microwave power suitably. The arrangement of the slightly modified [109] spectrometer as described in Fig. 4-10 may be used for this purpose. The null method may be used to measure the intensity of the

Γ-INTENSITY COEFFICIENTS FOR CHEMICAL ANALYSIS

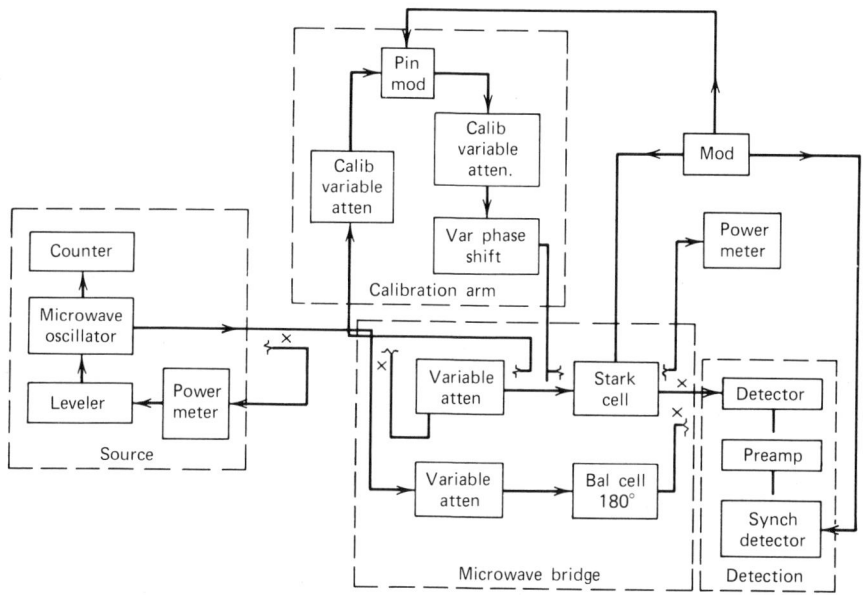

Fig. 4-10. Block diagram of modified HP Model 8460A spectrometer proposed by Curl. (With permission of Academic Press.)

line at its peak absorption frequency. The calibration arm signal is adjusted to be in phase with the microwave radiation entering the cell. Furthermore, the modulations of the two signals are identical. With no sample in the cell and no signal from the calibration arm, the null point of the synchronous detector is determined. The sample is introduced at a certain pressure. The phase of the calibrator arm signal is now adjusted by the phase shifter to 180° out of phase with the sample-absorption signal. The calibration arm attenuator is adjusted until the original null point is restored. The attenuator reading is a direct measure of the signal intensity. Power is then increased so as to maximize the signal attenuator setting, which is recorded. This attenuator setting is then a direct measure of the maximum signal, S_{max}.

A calibration curve of $10^{-D_p/20}$ versus p_p, where D_p is the attenuator reading in decibels corresponding to the maximum signal from the pure gas and p_p is the pressure of the pure gas, is plotted. The attenuator reading, D_M, in decibels corresponding to the maximum signal of the gas in the mixture is now determined. The partial pressure of the gas, p_M, corresponding to a value of D_M is read from the calibration curve. The total pressure, p_t, of the mixture is measured and the mole fraction of the gas in mixture is calculated from the ratio p_M/p_t. The procedure described appears to be simple and convenient; however, a calibration curve has to be generated for each special case.

Rinehart [110] has suggested a slight modification of this procedure to simplify the calibration procedures. The attenuators in the calibration arm were set so that the peak of calibration signal was near the peak of the frequency of the molecular signal. The calibration signal once prepared can be used for a number of runs. Rinehart defines a relative signal, S_{rel}, by the equation,

$$S_{rel} = \frac{\text{maximum signal from sample}}{\text{signal from calibration arm}} \qquad (4-22)$$

The signal, S_{rel}, is insensitive to fluctuations of the detection system sensitivity.

Two sets of plots are prepared: (1) S_{rel} against p for a reference sample containing known, X_2, mole fraction of the absorber and (2) S_{rel} against p for gaseous mixture, containing the absorber of unknown mole fraction, X. The slope of the plot for the gaseous mixture is given by the equation

$$(\text{slope})_1 = \frac{Rxp_2 - Rxp_1}{p_2 - p_1} = Rx \qquad (4-23)$$

where p_2 and p_1 are two different pressure values on the curve and R is the proportionality constant relating S_{rel} to xp (i.e., $S_{rel} = Rxp$.). The corresponding slope for the reference substance can be written simply as

$$(\text{slope})_2 = Rx_2 \qquad (4-24)$$

Γ-INTENSITY COEFFICIENTS FOR CHEMICAL ANALYSIS 119

The mole fraction, x, of the absorber in the mixture is now calculated from Eqs. (4-23) and (4-24) and is given by the relation

$$x = \frac{(slope)_1}{(\overline{slope})_1} \cdot x_2 \qquad (4-25)$$

Rinehart used least-squares techniques to obtain the slopes for the analysis of mixtures of vapors of CH_3OH and C_6H_6. He has determined the concentration of CH_3OH with an accuracy of about 0.1% by this procedure.

The procedure of measuring maximized signal under optimum power saturation to estimate the Γ-coefficient fails in the case of lines with considerable pressure broadening. Sufficient power may not be available to increase the value of ϕ to ϕ_{max} and the signal to S_{max}. However, methods developed by Harrington [38,39] and White [111] can be applied for using the Γ-coefficient under conditions of below power saturation.

As mentioned before, a disadvantage of the power-saturation technique for measuring the Γ-coefficient is the requirement of relatively high microwave power levels. The OCS data plotted in Fig. 4-9 show that a modest power of 3.6 mW is needed to saturate the OCS line at 2.6 Pa. The power required for saturation varies directly as the square of HWHH, $\Delta\nu$, and inversely as the square of dipole moment matrix elements, $|\mu_{ij}|^2$. The strong intensity and narrow line width of the OCS line favors power saturation at a low power level. Crable [105] has calculated the power required for saturation of transitions from a number of molecules at 2.6 Pa pressure (see Table 4-8). The data show that for most molecules the power required for saturation is perhaps higher than currently available in commercial spectrometers using BWOs as the source. A calibration curve may be required for each separate determination of the Γ-coefficient. The total independence of the Γ-coefficient from line width or other line-broadening effects is indeed a great advantage and is unmatched by other techniques.

TABLE 4-8. MICROWAVE POWER REQUIRED FOR OPTIMUM POWER SATURATION

Molecule	Frequency, MHz	Linewidth Parameter, (kHz/mTorr)	Power Required at 2.6 Pa mw
CH_2O	22,965.57	24.7	150
	24,068.35	20.1	130
SO_2	20,335.43	18.0	83
	24,083.46	17.3	140
CHF_3	20,697.73	33.3	
CH_3CF_3	20,740.53	49.8	90
CH_3COCH_3	21,169.95	52.1	300
	23,656.28	50.4	260

4.6. SOURCES OF UNCERTAINTY IN ANALYTICAL DETERMINATION BY MICROWAVE ROTATION SPECTROSCOPY

It would be appropriate to summarize briefly the experimental difficulties the analyst must overcome and the pitfalls to be avoided in applying the MRS technique. The problem will depend on the methods of analysis used. Major problem areas are discussed in the following paragraphs.

4.6.a. Relative Intensity Measurements

The microwave power level used to monitor two or three transitions from different components in a mixture is kept constant during the relative intensity measurements. This is normally done by keeping the d.c. output from the microwave detector at a chosen constant level.

The rotational transitions should be fully modulated for analytical purposes. This requires a proper selection of the transitions and the sample pressure so that a high Stark modulating field can be applied to the molecules.

Distortion of line shapes, due to undisplaced Stark lobes or neighboring Stark lobes and lines, should be minimized or taken into account in computation of parameters. Also, power saturation should be avoided in relative intensity measurements based on determination of Beer's law coefficients. Reflections and mismatch in the wave guide should be avoided by careful design and electronic noise, and pick-up should be minimized.

The absorption path length, L, is not in general equal to the effective path length ℓ, (i.e., $\ell = L \cdot \lambda_g/\lambda$) and accordingly a correction should be applied, particularly in the measurement of lines widely differing in frequencies.

The group velocity, v_g, for the velocity of microwave radiation in the wave guide should be used in place of c, the velocity of light in the basic equation [Eq. (1-28)] for α_ν.

The temperature of the microwave cell should be kept constant during the relative intensity measurements. Also, the microwave absorption cell should be conditioned with the sample gas for certain time periods before commencing measurements. This would minimize the drift in pressure of the component under observation due to adsorption on cell walls.

4.6.b. Absolute Intensity Measurements

Absolute intensity measurements are in general very difficult. The most important sources of errors for these measurements are: (1) spectrometer characteristics, (2) undermodulation, (3) overlapping from other lines, (4) saturation effects, (5) baseline uncertainty, (6) errors due to inability to calculate the effect of centrifugal distortion on dipole moment, and (7) error in measurement of dipole moments. When these sources of error can be minimized, absolute intensity measurements can be made with better than 1% accuracy.

4.7 Sensitivity, Accuracy, and Precision of Quantitative Analysis by Microwave Rotational Spectroscopy

The best sensitivity of a commercial Stark modulated microwave rotational spectrometer such as the HP Model 8460A is considered to be about 10^{-9} cm^{-1} minimum detectable absorption coefficient, with a S:N ratio of 2:1. Other special spectrometers discussed in this book have about an order of magnitude higher sensitivity. Since the stronger absorbing molecules have absorption coefficients of 10^{-4} to 10^{-5} cm^{-1} (in the frequency range of the conventional spectrometer, i.e., 8 to 40 GHz), the minimum detectable concentration for these species is in the range of 1 to 10 p.p.m. Larger molecules, which typically have many lines due to intramolecular splitting effects, rarely have lines with absorption coefficients larger than 10^{-6}. With few exceptions, molecules with molecular weight higher than 150 can only be detected in concentrations of 0.01% or higher. Of course, the general sensitivity of MRS can be increased by several orders of magnitude by using gas separators (permeable membranes) or preconcentrators. Such methods for extending sensitivity are discussed in Chapter 5.

The advent of automated spectrometers with computer-based methods of analysis has considerably improved the accuracy and precision of analytical determinations. The accuracy and precision of analytical determinations by MRS are rarely reported in the literature. However, a few reports quote the "figure of merit" of the determinations. Scharpen and Rasukolb [43] have reported on the analytical determination of deuteropropenes with a standard deviation of 2%. Rinehart [110], using slopes of straight-line plots of the maximum signal intensity with sample pressure, obtained mole fractions for concentrations of methanol in benzene with accuracies of 0.1%. This represents a considerable improvement over work done a decade earlier by Esbitt and Wilson [101], who measured peak intensity ratios of ^{12}C and ^{13}C species of enriched methyl formate to accuracies of 3%. Using absolute intensity-measurement techniques, Curl [109] has obtained values of free-energy functions with extremely high precision (e.g., 0.5%) in all cases. A precision of about 5% has been reported for trace-pollutant analysis in gas mixtures by the Gunn diode microwave spectrometer [56]. However, 10% is generally the level of accuracy expected in the determinations of mole percent of pollutants in air. The commercial instruments (both CSI and HP spectrometers) provide precisions of approximately 2% for α_{ν_0} measurements and 1.5% for $\Delta\nu$ measurements.

The precision of measurements for α_{int}, obtained by integrating the area under a line is expected to be similar to the precision for $\Delta\nu$ measurements (i.e., 1 to 2%). It is thus possible, in principle, to measure the absolute concentration of a species in a gas mixture to high precision without using calibration standards, although it would clearly be a difficult and painstaking task.

Microwave rotational spectrometry certainly is capable of providing accuracy, specificity, and repeatability to a degree competitive with any other single instrument for gas analysis. In addition, it offers the potential for absolute concentration measurement, a capability not possessed by any other popular instrumentation for gas analysis.

4.8. Type and Size of Sample and Sampling Consideration for Analysis by Microwave Rotational Spectroscopy

The main requirements of a sample for analysis by MRS are that the substance have a nonzero electric (or magnetic) dipole moment and sufficient vapor pressure at the temperature of the observation. The minimum pressure required for analysis is approximately 0.1 Pa. Most liquids and even some solids have sufficient vapor pressure at room temperature. Special high-temperature cells can be used for many substances to obtain the appropriate vapor pressure. The Stark absorption cells used on the commercial instruments usually can be heated to temperatures as high as 120°C without damage.

Very little sample is needed for analysis by MRS. The typical Stark cell is 2 m in length and has a volume of approximately 0.5 liter. This volume, filled to a pressure of 6.5 Pa at $T = 300$ K, requires only 1.5×10^{-6} mol of gas, or (for a sample of molecular weight of 100) only a few hundred micrograms of material.

Since the microwave radiation used for an analysis does not carry enough energy to decompose the sample, the analysis can be nondestructive, that is, the sample is essentially unaffected and can be recovered in its original state. However, because the absorption cell is made of metal and usually has a large surface:volume ratio, some problems with adsorption and chemical reactivity in the cell can arise for certain molecules. Some molecules, particularly ammonia, water, formaldehyde, and sulfur dioxide, are notorious for adsorption in microwave adsorption cells. As described in Chapter 5, these problems must be overcome when trace-quantity measurements are desired.

Chemical reactivity can present serious problems if the lifetime in the cell is not sufficient for observation of the spectrum. Special cells, such as an open space resonant cavity [55] or parallel plate Stark cells [15], have been designed specifically with large volume:surface ratios to overcome these problems. Sometimes reaction rates can be reduced by "conditioning" the cell, that is, filling the cell with successive samples at a somewhat high pressure. For very reactive species, a continuous flow system is desirable. Thus a fresh sample is pumped continuously through the cell, and the flow rate is adjusted to obtain the desired pressure and renewal rate. The reaction itself can also in many cases be reduced sufficiently by cooling the cell. As in most other analytical techniques, sampling problems can be a definite limitation to the utility of MRS for ultra trace detection; when these problems are solvable, MRS provides some very desirable features for qualitative and quantitative gas analysis.

CHAPTER

5

APPLICATIONS AND FUTURE POTENTIAL

The preceding chapters have provided detailed information about the MRS technique, emphasizing the features that qualify its analytical application. The prime requirements of an analytical instrument are, of course, its qualitative specificity for the compound or compounds being analyzed and the precision of the quantitative value measured by the instrument and assigned to the specific compound. Other properties such as stability, reliability, simplicity, and ruggedness are secondary but can rate along with the prime requirements, depending on the specific applications. It should also be noted that the cost of the analytical instrumentation can be most important for the broad application of any technique, including MRS. For example, low cost can be the sole reason for the high utilization of inferior techniques, whereas high cost can offset the desire for a superior measurement.

Microwave spectroscopy probably falls within the second category. Although its utilization is low, its applications are many. This chapter discusses the applications of MRS in which it has already provided results that can be directly compared with other methods. Other applications are discussed in which the technique shows promise and for which some preliminary work has already shown feasibility.

Some prognosis of the future potential of MRS is given in this chapter. This is mainly based on two conditions that the authors feel will change the current situation: (1) the realization of much simpler, more reliable, and relatively inexpensive instrumentation and (2) the future need in energy related systems for compound specific, on-line process control monitors. Either one or both of these could provide the impetus for wider use of MRS in the future.

5.1 REACTION STUDIES BY MICROWAVE ROTATIONAL SPECTROSCOPY

Techniques of microwave rotational spectroscopy have been applied to: (1) detection and characterization of transient species formed during the course of chemical reactions, as well as the final products of the reactions, (2) quantitative determination of the amounts of the products reactions, and (3) monitoring of the kinetics of chemical reactions, to elucidate the mechanism of reactions.

Microwave rotational spectroscopy can provide the information relating to the identity as well as the amount for each of the reaction products. This is a new kind of information for the catalyzed hydrogen-exchange reactions of unsaturated hydrocarbons (e.g., propene) and has never been obtained by other common techniques such as mass spectrometry, tracer study, IR spectroscopy, or NMR. The characterization and quantitative determination of the five propene-d_1 species, differing solely from one another in the location of the deuterium atom [43], discussed in detail in Chapter 4, is typical of such an application.

Extensive application of MRS for chemical reaction studies has been made by the Japanese since the mid-1960s to the present time. Morino and Hirota [112] gave a description of their technique as early as 1964 for determining the relative abundance of deuterated propylene in hydrogen-exchange reactions. Subsequent work made application of MRS to pyrolysis studies [113-115] (discussed in the next section), then to catalysis studies [116-119], which continues to the present time. The microwave technique has proven to be a most powerful tool for such studies, both as a selective analyzer of reaction products, and as an <u>in situ</u> technique for real time measurement of transient species [120].

Microwave spectroscopic techniques were used to investigate the mechanism of hydrogen-exchange reaction of propene [61,104,121,122] in the presence of a variety of catalysts. Associative mechanisms based on the formation of n-propyl or isopropyl reaction intermediates and dissociative mechanisms that assume formation of π-allyl intermediates have been proposed for the catalyzed hydrogen-exchange reactions of propene. The two types of proposed reaction mechanisms [120], as well as the projected rates of formation of products, are shown in Fig. 5-1. The D_2

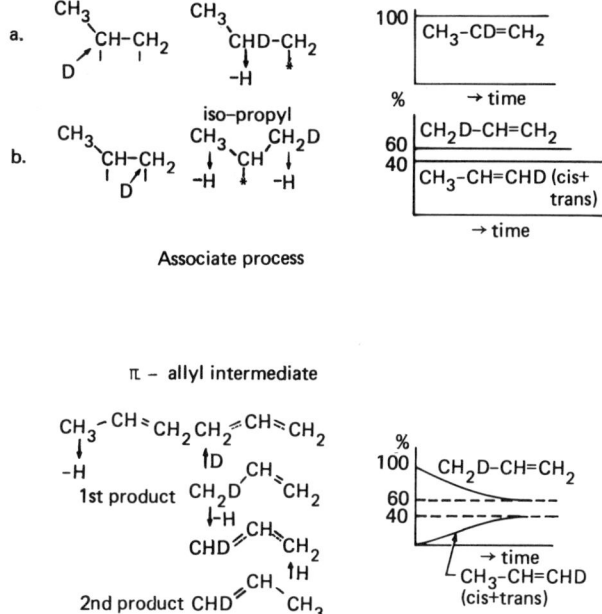

Fig. 5-1. Rates of formation of products from hydrogen-exchange reactions of propene.

molecules dissociate into D atoms. The propene molecule may open its double bond on adsorption at catalyst surfaces, and the D atom can add on to either of the two carbon atoms to produce n-propyl or isopropyl reaction intermediate. The reaction intermediate can go back to the parent propene molecule by eliminating a hydrogen atom. Formation of a single reaction product, $CH_3-CD=CH_2$, is suggestive of the proposed associative mechanism with the formation of n-propyl intermediate. The formation of $CH_2D-CH=CH_2$ as well as cis and trans $CH_3-CH=CHD$ in the ratio of 3:2, and the fact that the ratio does not change with time as the reaction progresses, suggest an associative mechanism based on isopropyl reaction intermediate.

The final ratio of two species, $CH_2D-CH=CH_2$ and $CH_3-CH=CHD$ (cis and trans) for the case of π-allyl path is projected to vary with time (Fig. 5-1), even though the final ratio, that is, of 3:2, is the same as for the case with isopropyl intermediate. Microwave rotational spectroscopy was used to determine the amount of monodeutero species as well as the rates of their formation. This procedure has been applied to reactions conducted over a number of catalysts, and the catalyzed hydrogen-exchange reactions were observed to conform to one of the three behaviors outlined in this paragraph. The results are summarized in Table 5-1. The reaction path diagnosed by the microwave spectroscopic study is indicated in Table 5-1 by specifying the reaction intermediate.

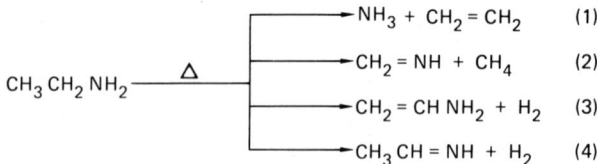

Stage I of Ethylamine Prolysis

$$CH_3CH_2NH_2 \xrightarrow{\Delta} \begin{cases} NH_3 + CH_2=CH_2 & (1) \\ CH_2=NH + CH_4 & (2) \\ CH_2=CHNH_2 + H_2 & (3) \\ CH_3CH=NH + H_2 & (4) \end{cases}$$

Stage II of Ethylamine Pyrolysis

$$CH_2=NH \xrightarrow{\Delta} HCN + H_2$$

$$CH_3CH=NH \xrightarrow{\Delta} \begin{cases} HCN + CH_4 & (5) \\ CH_3CN + H_2 & (6) \\ CH_2=C=NH + H_2 & (7) \end{cases}$$

$$CH_2=CHNH_2 \xrightarrow{\Delta} \begin{cases} HC\equiv CNH_2 + H_2 & (8) \\ NH_3 + HC\equiv CH & (9) \end{cases}$$

Fig. 5-2. Two-stage cracking pattern postulated for pyrolysis of ethylamine. (From Lovas, Clark, et al. [131], with permission.)

TABLE 5-1. CATALYZED HYDROGEN-EXCHANGE REACTIONS OF PROPENE WITH D_1

Intermediate	Catalyst	Reference
Isopropyl	D_2SO_4/D_3PO_4	122
π-Allyl	potassium-graphite intercalation compound	123
π-Allyl	ZnO	121
Isopropyl	D-p-toluenesulfonic acid/SiO_2	119
Isopropyl and n-propyl	Ni, Co-phthalocyanine-Na^+ complex	124

The MRS technique has also been used [125] to measure the distribution of propene-d_1 species formed by the hydrogen-exchange reaction over nickel and palladium catalyst. The distribution of propene-d_1 species found in the reaction products is given in Table 5-2. The results shown in Table 5-2 were interpreted by the investigators to indicate differences of catalytic activity between nickel and palladium; in other words: (1) CH-hydrogen is most exchangeable over Ni, (2) CH_3-hydrogen is slightly more exchangeable than the CH_2- and CH-hydrogen over Pd, and

TABLE 5-2. DISTRIBUTION OF PROPENE-D_1 SPECIES IN PRODUCTS OF NICKEL- AND PALLADIUM-CATALYZED REACTIONS[a]

Catalyst	$CH_2DCH=CH_2$ (%)	$CH_3-CD-CH_2$ (%)	cis $CH_3-CH=CHD$ (%)	trans $CH_3-CH-CHD$ (%)
Pd	59.2 ± 2.1	14.1 ± 1.3	13.7 ± 1.5	13.1 ± 1.3
Ni	22.5 ± 2.7	58.1 ± 5.1	9.9 ± 2.2	9.5 ± 2.1

[a]Source: Hironaka and Hirota [126].

(3) exchangeability of two hydrogen atoms at the CH_2 group is the same irrespective of the cis or trans configuration relative to the CH_3-group. This study was followed by another similar study [126] on the hydrogen exchange of propene in the presence of nickel and palladium catalysts.

Kim and Gwinn [127] also used microwave techniques in a qualitative manner to study the rearrangement of cyclobutyl cation (considered as an intermediate in the reaction of dried cyclobutanol with PCl_5 in the presence of D_2O) through a definite identification of the deuterated products.

5.2 PYROLYSIS AND OZONOLYSIS STUDIES

Microwave rotational spectroscopy has become popular within the past 10 years for use in identifying products from pyrolysis and ozonolysis. Here its shear power to spectrally distinguish the many different species produced, makes it an indispensable tool. For example, in studies of ozonolysis by isotopic labeling, the resolution of MRS readily permits unambiguous detection of different isotopic spectra. Since it is nondestructive, the analysis can be made without destroying the ozonide, in contrasts to isotopic mass-spectroscopic studies or chemical degradation to simpler species with subsequent mass-spectral analysis. Hence assumptions about fragmentation are not necessary.

The detection of products from pyrolized compounds by MRS has been facilitated by the designs of high-temperature cells [57,128], where the species can be observed spectroscopically in situ. Several laboratories have now constructed such systems and are currently heavily involved in pyrolysis reaction studies with this powerful new analytical tool.

One of the first thorough pyrolysis studies published in which MRS was used as the main analytical method was that by Saito [113], who studied the thermal decomposition of ethylene episulfoxide by observing intermediates during the reaction as well as analyzing the end products. His observations are shown in Table 5-3. From his data, Saito was able to show that ethylene episulfoxide decomposes on a hot surface to yield both ethylene and the intermediate, SO, both in their ground states, with the SO being in the triplet σ-state. The SO further decomposes

TABLE 5-3. COMPOSITION OF PYROLIZED ETHYLENE EPISULFOXIDE[a]

Molecule	Percentage	Method
$(CH_2)_2SO$	2	CRO
SO	~3	CRO
SO_2	20	CRO
S_2O	~24	CRO
H_2CO	0.8	CRO
$(CH_2)_2 S$	≤0.3	recorder
$(CH_2)_2 O$	≤0.04	recorder

[a]Source: Saito [113].

into SO_2, S_2O and sulfur. In later work, Saito and Wentrup [115] performed similar studies on the pyrolysis of N-sulfinylaniline and confirmed its dissociation into phenylinitrene and sulfur monoxide.

A considerable amount of work has been done more recently by a group at the National Bureau of Standards (NBS) in Washington, D.C. The first work reported from this group [129] involved the detection and identification of thioformalydehyde (H_2CS) from the pyrolysis of CH_3SSCH_3. This work has since had direct bearing on the successful finding of H_2CS in interstellar space.

By pyrolizing methyamine CH_3NH_2 in a 4-mm dia quartz tube with a 2-cm-long section heated very near to the melting point, methyleneimine (CH_2NH) was identified as one of the products flowing through the attached microwave spectrometer by the NBS group [130]. In addition, they were able to determine the structure of this transient species as well as an approximation of its lifetime (0.1-s half-life in their system). For these studies, the analytical cell was operated at room temperature and at a pressure of 1 to 3 Pa. The flow rate was estimated to be 0.1 g/h. Pyrolysis products were only observed for temperatures about 700°C at the quartz tube. Accurate structure determinations were made by comparison of spectra from five different isotopic derivatives of CH_3NH_2.

More recently the NBS group has been studying the pyrolysis of ethylamine ($CH_3CH_2NH_2$) [131]. In contrast to the simple decomposition of methylamine into CH_2NH, which subsequently decomposes into HCN, the pyrolysis of ethylamine can theoretically follow a complex two-stage cracking route that produces nine stable polar species and three nonpolar species. Both stages have been identified by the NBS group by way of the observation of the microwave rotational spectrum for six of the polar species. By piecing together information gained from the microwave spectra, the general cracking pattern for ethylamine (Fig. 5-2) was established. The first-stage cracking pattern was determined by direct observation of absorption lines assignable to each of the polar products indicated. The second-stage cracking pattern is somewhat more tentative in that products (7) and (8) were not identified. However, HCN, NH_3, and CH_3CN were all observed. The cracking seems to favor a simple unimolecular decomposition since signals from the species CH_3NH_2 and H_2NNH_2, expected to be formed by the recombination of the radicals CH_3 and NH_2, could not be observed. This is certainly a most lucid example of the power of MRS in qualitative analysis of complex mixtures and shows why there is a growing interest in MRS from oil-, chemical-, and energy-related industry as more is learned about this technique. Other pyrolysis studies with MRS relating to energy research is currently underway at universities in Denmark, Japan, Austrailia, West Germany, and Great Britain, as well as in the United States.

Ozonolysis studies have also recently become popular because of their importance to atmospheric chemistry. Again here, MRS is a powerful analytical tool since it can be used to identify new products as well as to quantify known reactants and products. Its additional ability to give precise information about structural conformations is of special importance in such studies.

Much of the recent work with MRS and ozonolysis has been done by a group at the University of Michigan. They have published work on ethylene ozonide- and oxygen-18-enriched ethylene ozonide [132] and propylene and trans 2-butene ozonides [133], among others.

In very recent work, Lovas and Suenram [134] have studied the ozone-olefin reaction believed to be important in photochemical smog generation and have successfully identified the ring form of dioxirane

(H_2COO) produced as an intermediate in the ozone-ethylene reaction. This product has never been observed from such reactions previously, but its finding supports recent theoretical work.

Examples given earlier for the application of MRS to the illucidation of reaction pathways show the versatility of the technique and suggest that it has more to offer than either mass or IR spectrometry for these kinds of studies. We do not find it unreasonable to predict that these qualities will be of special interest to researchers active in the field of reaction chemistry in the future. There is at least one known case where a researcher in this field, recognizing the power of MRS for his studies (but finding no commercial source for the instrumentation), constructed his own system (with a slight amount of consulting help). The individual was astonished that such a powerful analytical tool was not presently commercially available.

5.3 ISOTOPE-RATIO MEASUREMENTS

Microwave rotational spectroscopy has an almost unique inherent capability for resolving different isotopic derivatives of the same compound. This is clearly demonstrated by the example shown in Fig. 5-3 for an isotopic mixture of HDSe. Here the five major isotopic species of Se are clearly identified in a spectrum obtained at dry-ice temperature in a conventional Stark cell [135].

Early in its development MRS was recognized for its possibilities to determine isotope ratios. Very early work was done by Townes, Holden, et al. [136] to determine natural abundance ratios of ^{37}Cl to ^{35}Cl from relative intensity measurements of absorption lines of $^{35}ClCN$ and $^{37}ClCN$. They found this ratio to be 0.3 and later used MRS to measure the concentration of the radioactive isotope ^{36}Cl [137].

Probably one of the best early examples of quantitative analysis by peak-height measurements for determination of isotopic abundances is the work by Southern, Morgan, et al. [138]. These workers developed a method of calibration in their spectrometer that allowed them to determine the ^{15}N abundance in ammonia and ^{13}C abundance in cyanogen chloride. Calibration curves were made by measuring the peak height ratios of $^{15}NH_3$ and $^{14}NH_3$ for various prepared

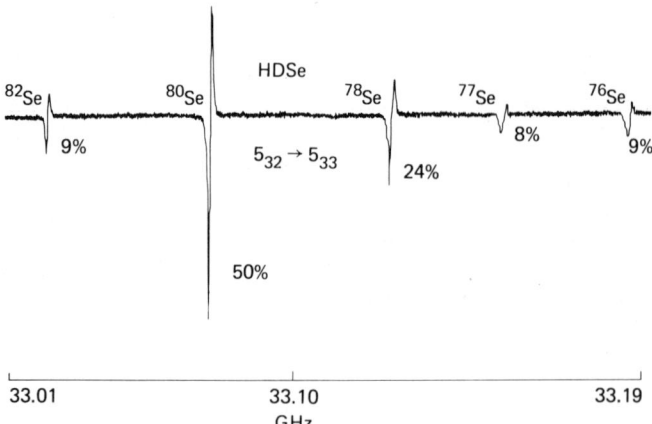

Fig. 5-3. Transition of asymmetric rotor HDSe. Spectral trace recorded from left to right extends approximately 170 MHz. The five major isotopic species of Se are clearly evident. Spectrum was taken at dry-ice temperature; nevertheless, HDSe decomposes in Stark cell. This is evident from decrease in line intensity of ^{76}Se compared with that of ^{82}Se. (From Jones and Cook [135], with permission.)

concentrations of ^{15}NH$_3$ and then dividing the ratio of peak heights of ^{15}NH$_3$ to ^{14}NH$_3$ in normal ammonia. The resultant values were plotted against concentrations of ^{15}NH$_3$. Then by measuring the ^{15}N:^{14}N ratio in an unknown sample, the concentration of ^{15}N was obtained from the calibration curve. For ^{15}N occurring in the range 0.38 to 4.5% their accuracy was within 3% of its concentration. For ^{13}C in the range of 1.1 to 10% the average error was less than 2% of its concentration. Their procedure required 10 to 15 min and involved 10 to 15 measurements per sample, although five measurements are believed to be sufficient for the accuracy obtained. Only 15 mmol of gas was required for the analysis. Another example of isotope-ratio measurements was the study of deuterated ammonia ratios in catalyzed NH$_3$-D$_2$ isotopic exchange reaction [139], where all of the deuteroammonia combinations could be measured by MRS.

Esbitt and Wilson [101] have determined the ratio of the abundance of ^{12}C to the ^{13}C species of enriched methyl formate to be 0.635 with a 2.8% standard deviation by measurements on the same transition in the two isotopic species. Their results were in good agreement with a mass-spectroscopic determination that gave 0.644 with a 3% uncertainty.

More recent applications of MRS for isotope analysis include unpublished work at the University of California, Lawrence Livermore Laboratory (UCLLL) to determine deuterated ammonia content in gas mixtures and a determination of chlorine isotope ratios in chlorosilane mixtures. In both of these applications the microwave spectrometer was set up to monitor the species continuously in a flow system to determine any change of the isotope-ratio balance due to the presence of potentially catalytic surfaces.

Microwave spectroscopy is probably being overlooked for many potential applications involving the need for isotopic selectivity, simply because of the lack of knowledge about its capabilities by potential users. One quite logical application would be its use for isotopic purity measurements for enrichment processes. It is indeed interesting to speculate on the impact of small, reliable, and inexpensive microwave instruments that could monitor isotope ratios in real time, be used for tracer studies, and so on. Several possible instruments of this type are described in a following section.

5.4 MIXTURE ANALYSIS BY MRS

Probably the most general application of MRS is qualitative and semiquantitative analysis of gas mixtures. Although limited in general to the analysis of low-molecular-weight compounds (mol.wt. <200), MRS can nicely compliment mass spectrometry and gas chromatography for complex mixture analysis of small polar molecules. Once armed with a comprehensive catalog of absorption frequencies and strengths for the range covered by his instrument, the analyst's job becomes quite easy. He can search the frequency range either manually or automatically (depending on the sophistication of his instrument) for signals at the frequencies of interest. The strong correlation between a compound and the microwave frequencies it absorbs assures the analyst of positive qualitative identification. He can increase his assurance with another or multiple checks for absorption at other preassigned frequencies. Having observed a sufficiently strong signal, any number of methods are available to obtain quantitative information about the specific compound being measured.

Following are a few examples of the application of MRS to mixture analysis. Note that most of these represent special analysis requirements and do not themselves emphasize the utility of the technique as a general analytical tool. Such general utilization is limited to a very few analytical laboratories since most microwave rotational spectrometers in current use are research oriented. An example of MRS used as a general analytical tool at the UCLLL is discussed.

A typical example of mixture analysis from the earlier literature is the work of Weber and Laidler [139] to determine concentrations of NH_3 in mixtures with deuterated ammonias. This work was done before the development of sensitive modulation-type spectrometers but still illustrated the unique capability of the technique to clearly distinguish slightly differing isotopic compounds in a mixture. Concentrations of NH_3 were determined by comparing the detector signals with and without the presence of the sample at the frequencies of strongest absorption for ammonia. The concentrations thus determined were considered accurate to within 0.5%.

After the development of very sensitive wide-band commercial spectrometers in the 1960s, examples of mixture analysis increased. Hewlett-Packard, for example showed in one of its technical reports [140] the ease with which SO_2 could be identified in a mixture with ethyl and methyl mercaptans and in another example [141], the detection of specific aldehydes in a mixture. A portion of the spectrum obtained for that mixture is shown in Fig. 5-4. It should be noted that any of these species can be detected without interference from the others simply by tuning the source frequency to coincide with the peak of their respective absorptions.

Another demonstration of mixture analysis was given by Funkhouser, Armstrong, et al. [142], for acetone, methanol, and freon in nitrogen. In this work several techniques for relative intensity measurements were compared in the quantitative determination of the selected compounds: (1) maximum signal by saturation method (see Section 4.5), (2) a constant

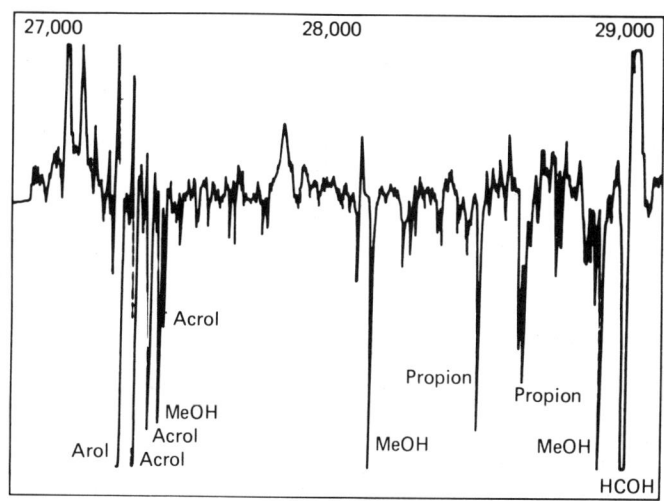

Fig. 5-4. Aldehyde mixture for frequency region 27 to 29 GHz. (From Hewlett Packard Application Note 840-5, with permission.)

source power slightly below saturation with measurement of just peak height, and (3) same as (2) but integrating the area under the recorded peak. All three of these measuring techniques gave good quantitative results, but (3) gave the most linear response over the widest range of concentrations and total pressures (0 to 10%; and 4 to 13 Pa). The reproducibility of this method is illustrated in Fig. 5-5, which shows three separate scans of an acetone-absorption line. The results of this mixture analysis agree very closely with mass spectrometry on the same samples, and the time required for the analysis is somewhat less than that required for mass spectrometry (∼5 to 10 min).

Using an automated HP Model 8460 spectrometer, White at NASA Langley Research Center has demonstrated analysis of a number of mixtures. An example of an analysis of a mixture including carbonyl sulfide, ethanol, methanol, formaldehyde, and fluorobenzine is described in one of White's reports [42]. Under computer control, White's spectrometer operates basically

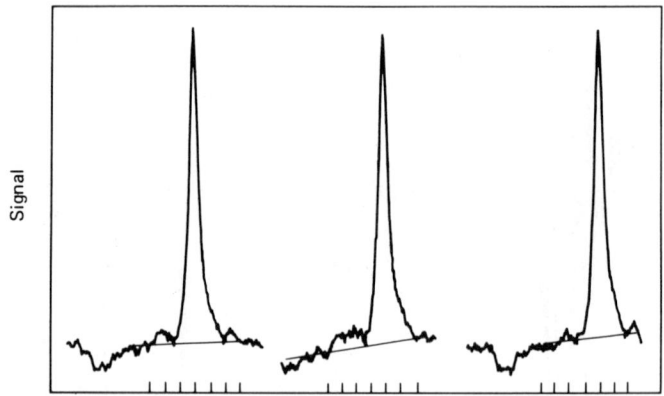

Fig. 5-5. Reproducibility of scans for the 36,934 MHz line of acetone. (From Funkhouser, Armstrong, et al. [142], with permission.)

in three modes during a search operation in an unknown mixture. It begins by scanning at a rapid rate across the specified frequency range, with a fixed gain setting. Any signal detected as not being noise (i.e., above a threshold value) causes a shift to the second mode of operation. The scan rate slows down, the oscillator frequency is stopped at the approximate peak of the detected line, and the gain is automatically adjusted to keep a strong signal indication on the output meter. The final measurement mode consists of taking a series of data points spaced every 0.01 MHz across the line peak, using digital filtering and signal averaging as appropriate. These data are used to calculate the center frequency, the line width, and the intensity coefficient of the line, all of which are used to determine the concentration of the species producing the line. The search then continues. The final qualitative analysis checks the strongest lines found against those stored in the computer that identify specific compounds. Figure 5-6 shows a computer printout for an automatic analysis run giving concentrations obtained (with tolerances) for each of the major components found in a mixture. Such an analysis method would typically take about

```
SET POWER ATTEN. TO 17
SENSITIVITY? -80

LOW SIDE OVERLAP OF 27478    LINE
UPPER LIMIT IS    6.7        PPM OF AMMONIA

BASED ON 3 LINES OF 3    CHECKED, APPROXIMATE CONCENTRATION
IS    24    + OR -    .4        PERCENT OF CARBONYL SULFIDE

UPPER LIMIT IS   12   PPM OF 1,2-EPOXYETHANE

HIGH SIDE OVERLAP OF 35298.5   LINE
LOW SIDE OVERLAP OF 36566.3    LINE
  2    LINES MATCH OUT OF 6    CHECKED FOR ETHANOL

LOW SIDE OVERLAP OF 35520.6   LINE
LOW SIDE OVERLAP OF 39231.    LINE
BASED ON 3 LINES OF 9    CHECKED, APPROXIMATE CONCENTRATION
IS   .4         + OR -   .1        PERCENT OF FLUOROBENZENE

UPPER LIMIT IS 83.1       PPM OF FLUOROETHYLENE

BASED ON 3 LINES OF 3    CHECKED, APPROXIMATE CONCENTRATION
IS   .8         + OR -   0        PERCENT OF FORMALDEHYDE

HIGH SIDE OVERLAP OF 38711.8    LINE
  1    LINES MATCH OUT OF 11    CHECKED FOR FURAN

BASED ON 3 LINES OF 3    CHECKED, APPROXIMATE CONCENTRATION
IS    7.2      + OR -   .4        PERCENT OF METHANOL

BASED ON 3 LINES OF 5    CHECKED, APPROXIMATE CONCENTRATION
IS    111   + OR -  27.3       PPM OF SULFUR DIOXIDE

READY
```

Fig. 5-6. Sample computer output for automatic analysis run. Concentrations are only approximate since only γ_0 was available for intensity comparisons. From White [42], with permission.)

2 h. White has used this automated spectrometer to accumulate data for the most comprehensive line-listing catalog for pure gases available today [36].

A considerable reduction in analysis time can be obtained in an automated spectrometer if the line frequencies and calibration data can be stored in the computer. Such is the case for the spectrometer

currently operating as a routine analytical system at
the UCLLL. A block diagram of the system is shown in
Fig. 5-7. Note that both the frequency and the input
power to the cell can be controlled during normal
operation. To prepare the system for automated analysis, the two strongest absorption lines for each cataloged gas are stored on a disk memory along with
identifying number used as a cross-reference. Table
5-4, for example, shows a partial listing of the
strongest absorption lines in the 26.5 to 40 GHz frequency range for over 100 different species; the 8-
digit numbers are the frequencies in kilohertz. Also
stored on the disk memory is the slope of a linear
calibration curve determined for each of the pure
gases. Such a calibration is obtained as follows:

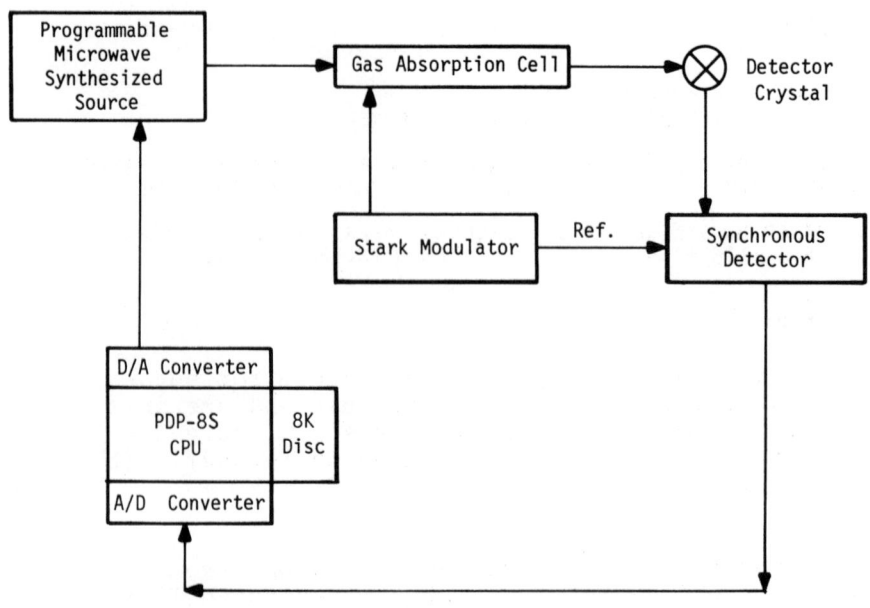

Fig. 5-7. Block diagram of automated analytical MRS spectrometer. Both frequency and
input power to cell are controlled during
normal operation.

TABLE 5-4. PARTIAL LISTING OF GASES STORED ON
DISK MEMORY FOR AUTOMATED ANALYSIS

Identifying Number	Chemical Name	First Strongest Frequency (kHz)	Second Strongest Frequency (kHz)
1	Acetaldehyde	38,512,109	38,506,074
2	Acetic acid	36,419,330	36,065,390
3	Acetone	38,349,475	36,933,770
4	Acetonitrile	36,795,529	36,794,358
5	Acrolein	38,847,733	37,904,867
6	Acrylaldehyde	35,578,110	36,435,980

The pure gas is admitted to the cell and the pressure is noted. The spectrometer frequency is then made to coincide with the peak of the strongest line, and the microwave power is automatically adjusted until the peak signal reaches a maximum. (Note: The crystal detector is maintained as a linear detector throughout). The pressure of the pure gas is changed and the procedure repeated until at least five data points are obtained. These data are then least-squares fit to a straight line whose slope is noted and stored for future use. Figure 5-8 shows typical calibration curves obtained in this manner.

The spectrometer can be operated either in a qualitative mode only or in both qualitative and quantitative, with the difference as a time savings by a factor of at least 2 for qualitative only. Typical operations for the qualitative mode would be as shown in the flow diagram in Fig. 5-9. After the sample is injected into the absorption cell, the computer program is put into operation, and the frequency is synthesized for the strongest absorbing line of compound No. 1. After about a second of signal averaging at the frequency, if a S:N ratio of 3:1 is observed, a second frequency is synthesized that corresponds to the next strongest absorption line for that compound. The signal is averaged for 2 s then if a S:N ratio of 2:1 is observed, the number of the compound is typed

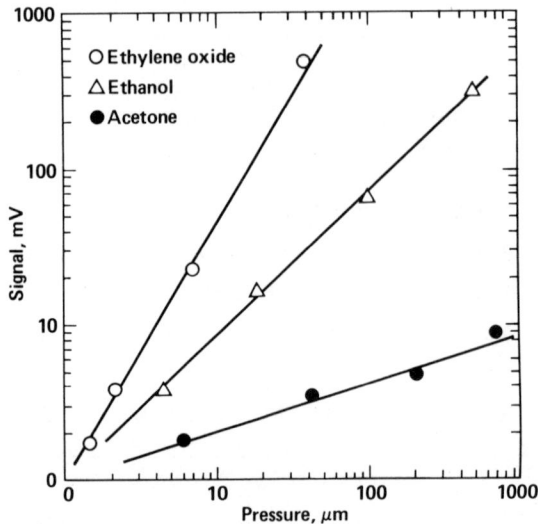

Fig. 5-8. Typical calibration curves determined by peak-saturation method for use in automated analytical spectrometer.

out indicating a positive identification. The process is repeated for the next compound in the list stored in the computer. If a signal is not observed for the strongest absorption frequency, the redundancy check is not made and the program moves directly to the next compound on the list. Typically, an unknown mixture can be searched for the presence of 120 compounds in less than 3 min. Table 5-5 shows a typical printout of a search made under computer control. This output is from an analysis of oil-shale retort residues made at UCLLL as part of the analytical characterization of the toxic components of such residues. A slightly longer operation is required to obtain a both qualitative and quantitative analysis of a mixture. After the computer makes the positive redundancy check, it again synthesizes the frequency of the strongest line. The microwave power is automatically increased incrementally while the signal is monitored for a maximum value. The maximum signal value is then used with the slope determined from prior calibration of the pure gas to calculate the concentrations or mole fraction

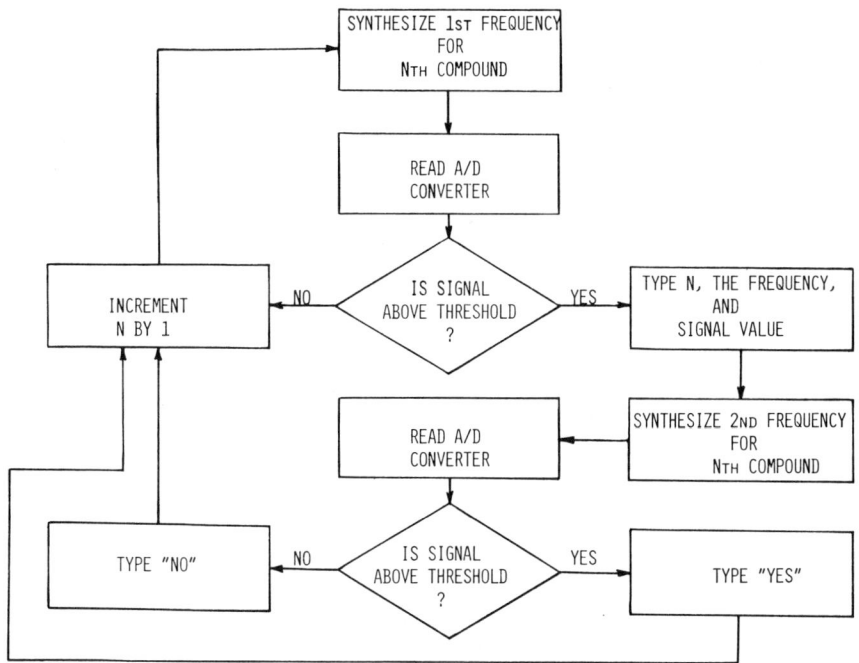

Fig. 5-9. Flow diagram showing algorithm used in automated analytical spectrometer.

of the species. Figure 5-10 shows typical data for a qualitative and quantitative analysis of a mixture of chlorofluorohydrocarbons by the manner just described.

Most of the examples of mixture analysis by MRS that have appeared in the literature are very specialized cases, where other techniques have failed initial attempts. Microwave rotational spectroscopy has not yet truly become a "routine" analytical instrument, with the possible exception of its current use at the UCLLL, where it is used to compliment the other gas-analysis methods. At the UCLLL this technique is usually the first choice for a qualitative analysis of an unknown mixture because of its speed. However, for precise quantitative measurements the time required is comparable with gas chromatography and mass

TABLE 5-5. RETORT TAR ANALYSIS BY MRS:
GO
:1 :1 :94
:10000
:.067

Gas No.	Frequency	Sig. Strength	Check	Name
4	36795529	2.571	yes	acetonitrile
9	27477934	1.011	yes	ammonia
16	36488818	0.073	no	carbonyl sulfide
26	38587900	0.199	yes	1-chloro-2-propanone
34	36417231	0.212	yes	ethanol
35	39296970	0.372	yes	ethyl amine
40	28974805	0.416	yes	formaldehyde
45	36597090	0.089	no	methanethiol
53	37699500	0.088	no	methylthionyl-amine
55	28741200	0.114	yes	phenol
64	35722270	0.077	no	propionitrile
81	39653535	0.119	no	ethanethiol
85	38908278	0.147	yes	methanamine
93	37238217	0.276	no	thiirane

spectrometry. Its major limitation as a "routine" analytical mixture analyzer is believed to be its lack of sensitivity for high molecular weight (mol.wt. ≥ 200) compounds.

Gas #	Frequency (MHz)	% Volume	Name
18	39,860.2	0.7	Chlorobenzene
20	34,550.1	7.1	Chlorodifluoromethane
23	34,405.9	<0.01	Chloroethylene
27	40,028.5	3.1	Chlorotrifluoromethane
28	36,494.9	3.0	Dichlorodifluoromethane
38	35,902.9	2.9	Fluorobenzene
39	37,991.2	<0.01	Fluoroethylene
76	39,452.4	0.08	Trichlorofluoromethane
		Remainder N_2	

Fig. 5-10. Quantitative analysis of mixture of chlorofluorohydrocarbons by automated MRS

5.5 AIR-POLLUTION MEASUREMENTS

Microwave rotational spectroscopy has been of interest for pollution measurements particularly because of its high resolution and excellent specificity. The gases and vapors typically considered pollutants, such as CO, NO, NO_2, and SO_2, all have permanent dipole moments and are small molecules; they are the most amenable to sensitive detection by MRS. In the course of the past 10 years, several investigations have been carried out to evaluate MRS pollutant measurements. Some of these were paper studies [143, 145], some were laboratory tests [56,105,146], and others were actually brought to practice with field-operable hardware [147-149]. The most pertinent of these studies are elaborated here.

It is possible to summarize the overall results of all of the studies to the present date by stating the following:

1. Microwave rotational spectroscopy does not have general applicability for air-pollution monitoring, mainly because it lacks the inherent sensitivity to detect the parts-per-billion levels. However, it does have general applicability for air monitoring where the limits of detection are parts-per-million or above, such as occupational environments.

2. Special-purpose spectrometers, with a means to concentrate the sample, are feasible alternatives to other methods, even at the parts-per-billion detection levels.

3. Microwave rotational spectroscopy compliments measurements made by other gas analyzers for air-pollution samples when the samples are returned to the laboratory for analysis.

4. Sampling problems, such as loss of sample to the walls of the inlet and absorption cell, chemical reactions, and so on, represent a major limitation of MRS for ultratrace gas detection.

The authors believe that the special purpose MRS spectrometers, like those developed for ammonia detection (see later section on portable monitors), can become cost competitive with other air-pollution monitors and thus may become popular in the near future.

Crable [105] has done an extensive study of MRS for air-pollution measurements. The compounds included in his study were SO_2, NH_3, CH_2O, NO_2, CH_3SH, and acetone. These compounds covered a range of types and problems of potential interest in air-pollution studies. Crable compared the saturation method for qualitative analysis with both the line-area and line-height methods. All methods gave linear response to concentration changes over some concentration range, with the saturation technique applicable to the largest concentration range, the line area, the second largest range, and the line-height method only over a considerably smaller range. The line-area method has an advantage over the saturation method because typically available microwave oscillators have insufficient power to saturate higher-pressure gas samples. However, the line-area measurements are more tedious. Otherwise, Crable found these two methods to give equivalent results.

Throughout his work Crable acknowledges many problems with sample handling, particularly adsorption problems. He points out, however, that the problems encountered in the microwave cells are even worse in the mass spectrometers. He concludes that the adsorption problem is a major limitation to the technique as a pollutant analyzer.

Crable's general conclusions are that MRS has limited sensitivity for most of the pollutant studies; specificity was very good as expected, and sampling problems are difficult to solve. He finally concludes that "for air pollution work, a microwave spectrometer should be considered as a very specialized analytical tool to be used only in those specialized cases in which its unique identification ability can be utilized."

Recognizing the limitation of poor sensitivity of MRS for pollution measurements, one of the authors investigated the use of resonant cavity absorption cells and dimethyl silicone membranes [56] to increase the sensitivity. The resonant cavity cell achieves greater sensitivity because it increases the effective absorption path length through the gas (up to many meters). Gas-permeable membranes effectively enrich the low-pressure sample in the species (usually the more polar ones) that permeate the membrane more rapidly, under a high-pressure driving force. The combination of these two devices increases sensitivity for most pollutant compounds (detectable by MRS) by two to three orders of magnitude. Thus the microwave technique, at least for special-purpose gas monitoring, can be used to detect pollutants down to levels of 10 p.p.b.; even lower detectability is obtainable with bulk preconcentration techniques.

A study was made to determine the effect of the membrane separator to enhance sensitivity to detect the polutants NO_2 and SO_2. The peak response of the cavity spectrometer was measured for each sample of a pollutant in air mixture. Plots of peak spectrometer signal versus the estimated concentration of NO_2 and SO_2 in the air samples were made with and without the use of membrane separators. The plots shown in Fig. 5-11 were found to be linear and reproducible within 5%. Zeeman (magnetic field) modulation was used for NO_2, and Stark (electric field) modulation was used for SO_2. The minimum detectable concentration, defined as the concentration in the air sample for a 2:1 S:N ratio, as found in this study are given in Table 5-6. The typical enrichment factors of over 100:1 result in a substantial increase in sensitivity for these pollutants.

The techniques described here have been embodied in prototype instruments developed for the Environmental Protection Agency (formaldehyde monitor) and the California Air Resources Board (ammonia monitor). These instruments are described in more detail in Section 5-9. Another instrument for detecting ammonia

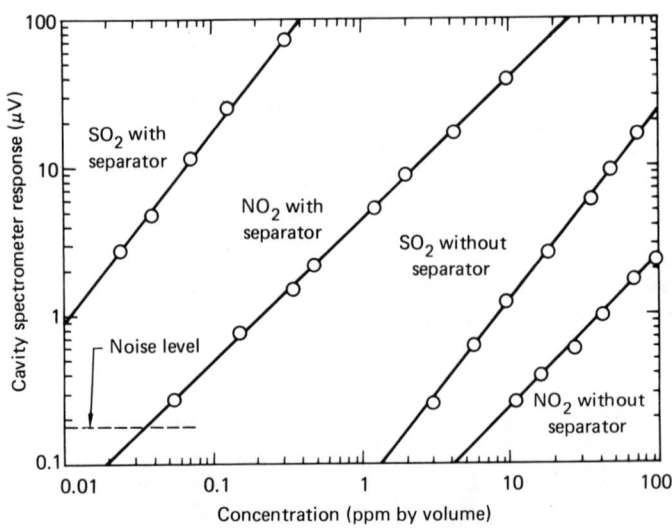

Fig. 5-11. Plots of signals for NO_2 and SO_2 in helium, with and without membrane separator at inlet of microwave spectrometer.

TABLE 5-6 MINIMUM DETECTABLE (S:N ~ 2) CONCENTRATION OF NO_2 AND SO_2 IN HELIUM GAS BY USE OF A GUNN EFFECT DIODE CAVITY SPECTROMETER

Contaminant	Membrane Separator Used	Without Use of Membrane Separator
NO_2	55 p.p.b.	11 p.p.m.
SO_2	25 p.p.b. (S:N ratio 10:1)	3 p.p.m.

is presently in field operation in the Imperial Valley of California; it is the only known field-operable microwave spectrometer for pollutant studies.

One of the few examples of actual pollution measurments made by MRS is the work done by Rinehart and Fletcher [150] to measure SO_2 concentrations in Wyoming near open air "flares" from crude-oil process plants. The method of slope ratios was used. The relative SO_2 signal from an HP Model 8460A spectrometer, S_{rel}, of a typical sample is defined as the ratio of molecular signal to the reference signal from the calibration arm of the spectrometer. A calibration curve was generated by plotting the values of relative signal, S_{rel}, against the partial pressure of SO_2 gas in the air samples. The linear plot showed negligible scatter of points. The SO_2 was separated from nitrogen, oxygen, water vapor, CO_2, and so on by using fractional distillation and condensation at low temperatures. The signal, S_{rel}, as well as the total pressure of the enriched gaseous sample was then measured. A comparison of these data with the calibration curve gave the SO_2 content in air. An analysis of laboratory air by this method gave values near 5 p.p.b. Samples of air collected in the vicinity of the crude-oil-processing sites were analyzed; those taken 1/4 mi from the site show SO_2 levels of about 28 ± 5 p.p.m., whereas those samples taken 3 mi from the site show about 6 ± 2 p.p.m. Rinehart points out that the H_2S present near the sites in reasonably high quantities does not interfere with the SO_2 measurement, and in fact can be independently measured by MRS.

White [148], in some unpublished work, has used his computer controlled MRS spectrometer to analyze gases trapped on chromosorb material. In measurements of laboratory samples he can detect quantities of organics (from fresh paints, paint thinners, etc.) down to 1-p.p.b. levels. White has also analyzed air from the Chesapeake Bay by passing up to 70 liters of air over chromosorb material, and then returning it to the laboratory for analysis of the trapped species. Isopropanol and methanol were detected from these traps at levels from between 13 to 30 p.p.b. Traces of sulfur compounds (SO_2 and COS) and acrylonitrile were also found. In general, White found difficulties with trap contamination by observing signals from aldehydes in all traps before the trap had sampled air.

Specialized spectrometers seem to be most applicable to air-pollution measurements, particularly as selective monitors. This technique has been favorably reviewed as compared with others [151-153] but has not yet been accepted for the measurement of any common pollutants.

As mentioned earlier, the MRS technique is particularly suited to air measurements where the detectability limits are less stringent than for atmospheric air. An example is the quality of air in working environments where many solvents are used in high concentrations. Threshold limit values for exposure are set by the Occupational Safety and Health Agency. These limits are usually above the 1-p.p.m. levels. With the preconcentration by membrane separators these levels are easily achievable by MRS. A prototype monitor for 10 selected solvent species is currently being developed by the UCLLL for the National Institute for Occupational Safety and Health (NIOSH) [154]. The instrument will be portable and will have the option of easily adding the capability for detecting other species as desired. When this instrument reaches the users, it is very likely to impress them with the selectivity attainable for air monitoring. It is reasonable to expect the technique to gain popularity from demonstrations of reliable and highly selective monitoring instrumentation such as this.

5.6 OTHER APPLICATIONS

5.6.a. Detectors for Gases in/from Solutions

Microwave rotational spectroscopy can be a very useful method for analyzing solutions that can be easily vaporized and for gases suspended in solutions. For those liquids having sufficiently high vapor pressure, the analysis with a conventional spectrometer is easily facilitated by injecting microliter quantities of the solution into the vacuum of the absorption cell through a rubber membrane. The subsequent analysis is accomplished in the same manner as for gas analysis.

Srinivasan [155] has recently described a cavity spectrometer that has possibilities for the determinations of ammonia in blood serum. His work shows that the cavity spectrometer can be used to give a linear output with concentration of ammonia, but its sensitivity seemed to fall far short of expectations. Concentrations of NH_3 in methanol and water as small as

100 p.p.m. were detected; 1-p.p.m. limits were sought. It is not clear from the published work why the instrument described lacked the expected sensitivity. A similar design (but at a higher frequency than the instrument described by Srinivasan) developed at the UCLLL, described in the next section, shows considerably higher sensitivity for ammonia (even without a gas-permeable membrane separator).

A rather novel sampling approach was used by Morrison [156] to determine ethanol content in water solutions. He used a small probe, which had a dimethyl-silicone gas-permeable membrane attached, placed directly in the solution. The vacuum pump for the spectrometer absorption cell pumped the permeated gases directly through the cell in a continuous flow for the analysis. A prime advantage of this technique is the elimination of adsorption problems normally observed with static measurements.

A similar application has been published by Aleksandrov and Tysovskii [157]. Also using a flow system and a broad-banded microwave spectrometer, they measured relative intensities to determine the concentrations of ethanol and isopropanol in gasoline. An average experimental error of 10% was obtained. The success of MRS in this determination is significant because analysis of this mixture by other methods is difficult.

5.6.b. Engine-exhaust Analysis

Two concepts are feasible for the application of MRS to automobile-exhaust analysis: The technique can be used for qualitative and quantitative analysis of mixtures that can be brought to the instrument for analysis and (2) microwave instruments can be designed to monitor the compound directly in the exhaust line itself.

Rinehart [149] has made some measurements of formaldehyde content in car exhaust. His procedure was to capture a large volume of exhaust gas in a bag, transfer the bag to his laboratory, then quickly freeze out all condensibles using a liquid-nitrogen trap. The remainder of the gas was pumped away. Then the condensibles were revaporized directly into the microwave absorption cell. Poor accuracy was obtained in these measurements because of large uncertainties due to the fate of H_2CO on the walls of the capture bag and the waveguide (as well as possible photochemical degradation). However, concentrations from 0.1 to 1 p.p.m. were typically measured. Rinehart [158] describes another method of integrating an absorption line to enhance the S:N ratio for detecting formaldehyde in automobile exhaust.

A specialized resonant cavity type spectrometer has been proposed by Easley [159] as a formaldehyde monitor from automobile exhaust. The design consists of a Gunn effect oscillator as the microwave source, a low-Q resonant cavity as the absorption cell, and balanced bridge detection plus Stark modulation. A prototype was constructed, and laboratory tests indicated sensitivities in the 100-p.p.b. range for formaldehyde. As far as is known, the instrument was never taken to practice for real exhaust analysis.

Another cavity-type spectrometer developed by Uehara and Ijuuin [160] was actually used to measure various aldehydes from car exhaust. They were able to detect formaldehyde at 10-p.p.m. levels from automobile exhaust and found acrolien and other aldehydes present at the 20 to 60-p.p.m. range. This is the only known direct application of MRS to automobile-exhaust analysis, but it appears that this technique may become important for such analysis with new emphasis on alternate fuel research involving methanol and other light alcohols. Aldehydes are the major pollutant expected from these fuels, but very few techniques are available for their analysis in mixtures, and none are known to have the real-time monitoring capability of MRS.

5.6.c. Cigarette-smoke Analysis

Several MRS spectroscopists have reported making analyses of cigarette smoke; much of this being done without formal justification, but rather out of curiosity. Harrington [161] observed at least eight species from a single puff of tobacco smoke: (1) CH_3CN, (2) CH_2CHCHO, (3) $(CH_3)_2CO$, (4) CH_3OH, (5) $HCHO$, (7) HCN, and (8) CH_3CHO. The signals were sufficiently strong for a quantitative analysis, but no attempt was made to do so.

Rinehart [162] analyzed both cigarette smoke and the gases given off from a cigarette filter. He found essentially the same species as Harrington but noted the very large amount of HCN in the analysis. Harris and Jones [163] have also analyzed cigarette smoke by MRS with similar results.

Although none of the above work has been published, the power of the technique for at least a qualitative analysis of very complex mixtures is clearly illustrated. Used in conjunction with other analytical tools, MRS can not only compliment them in qualitative and quantitative application, but it also provides additional measurement capability, such as structural determinations and dipole moments.

5.6.d. Future Potential

It is not easy to speculate on the future of MRS as an analytical technique. There were times in the past 20 years or so when it appeared as if it might take its place among the other established gas-analysis techniques, but this has not happened. This is typified by projections made in recent years by a trade journal [164] that showed the growth potential of MRS to be third highest (percentwise) among 16 analytical techniques. Since that time, however, the situation has drastically changed; there no longer are any MRS instruments commercially sold in the United States. The circumstances responsible for its lack of growth and popularity are many. Right from its inception, the instrumentation was expensive. The electronic sophistication was also a drawback. Probably for this reason, inexpensive and easy to use instrumentation for classroom demonstrations was unavailable in the late 1940s and 1950s. Thus the technique was essentially missing from the typical analytical chemists curriculum through its early years. Actually, rotational molecular spectroscopy theory is still not nearly as popular a topic as vibrational molecular theory even though it has similar intrinsic information about molecular structure.

The 1960s brought an increasing interest in rotational molecular spectroscopy, particularly as a research tool. Several companies started development of microwave spectrometers. The developers included three American companies, Sperry, Tracerlab, and Hewlett Packard, and a Japanese firm, OKI. None of these companies, however, produced an inexpensive instrument, although it is believed that the technology was surely available at that time to manufacture them. Consequently, a rather large investment was needed, both for instrumentation and manpower, by any potential user for analytical purposes. Few industries were willing to make such an investment in an unproven, and generally unpopular, technique. The authors believe that the market existed then, and still does, for inexpensive, special-purpose instrumentation. Technology advances, particularly in solid-state sources, better detectors, and microprocessors, certainly now opens the possibility for inexpensive spectrometers. We describe some proven designs for potentially inexpensive systems and speculate on others in this chapter.

If inexpensive, special-purpose spectrometers can be made and marketed, we believe that MRS will see another rise in popularity in the next few years. The major impact could come as an on-line process monitor for the process-control industry or as portable gas monitors. The power of a combined gas-chromatograph/microwave spectrometer for analysis of complex chemical systems may also become evident. We discuss these spectrometers in detail in the following sections.

5.7 INEXPENSIVE ANALYTICAL SPECTROMETERS

Microwave rotational spectrometers are basically simple. The elements of the spectrometer are: (1) a slightly tunable microwave oscillator, (2) a length of wave guide that is appropriately vacuum sealed, (3) a diode detector, and (4) a vacuum pump. Sufficiently strong absorbing lines such as those from ammonia gas can be observed directly as sharp "dips" on an oscilloscope whose time base is synchronized with the frequency sweep of the source. This technique is called "video detection," and though quite simple, it is not particularly sensitive; it is mentioned here to show the simplicity of the MRS technique. The hardware for this instrumentation, including the power supplies and vacuum pump but excluding the oscilloscope, would cost about $500 at current prices if assembled from commercially available devices. However, unless the oscillator has a tuning range of at least 1 GHz, its usefulness is quite limited for general spectrometry; it would essentially only be useful for single-gas detection or at most, a few gases. The commercial spectrometers discussed in earlier chapters generally have oscillators with a wide tuning range and stabilization, but they are by no means inexpensive. Several types of inexpensive spectrometers are described here. Such designs have been made to work in the laboratory case, but none of these are commercially available at this time.

INEXPENSIVE ANALYTICAL SPECTROMETERS

A spectrometer designed to detect only one gas can be reasonably simple and inexpensive because the oscillator need not have a large tuning range. One design for a simple spectrometer is shown in Fig. 5-12. Gunn effect diodes are currently available and prepackaged in a mount for a cost of only $25 each in quantity purchases. A typical one of these oscillators has an output power of 5 to 10 mW and tuning range of about 30 MHz (tunable by varying the bias voltage) at any center frequency of choice between 8 and 12 GHz. A good low-noise silicon crystal detector diode (1N23B or equivalent for the 8 to 12-GHz band) can be purchased at a cost of about $10 each; a tunable mount for the diode would cost about $80. The absorption cell can be purchased commercially as a length of copper wave guide, with o-ring seals in the flanges, at a cost of $10 per foot. Vacuum windows are needed on each end of the waveguide. Mylar or mica sheets serve as strong material with good microwave transmission properties; either of these can be obtained at low cost. The least expensive vacuum pump with a sufficient base pressure for MRS ($\sim$0.6 Pa) costs about $120. The power supply for the Gunn effect diode oscillator should provide a regulated current of about 0.5 A at 6 V, and should be voltage programmable or sweepable. Such supplies are commercially available at a cost of $130. The detected signal can be displayed on a microvoltmeter (cost $\sim$$200). However, an

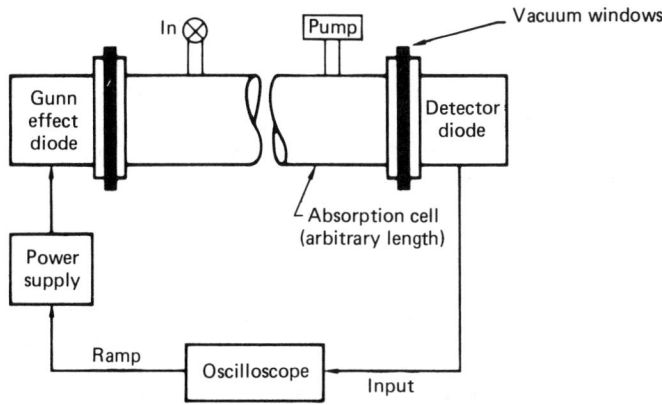

Fig. 5-12. Simplified spectrometer designed to detect a single gas.

oscilloscope, with a high-gain preamplifier, is preferred but of course costs considerably more than a voltmeter.

Such hardware makes up a fully operational spectrometer to monitor a single gas. Its total cost would be about $600. Such an instrument would require some modification to the electronics and wave guide to be reasonably sensitive. Stark modulation and synchronous detection would enhance the S:N ratio considerably. The additional electronics would be slightly more costly than the d.c. microvoltmeter (∼$250). To be useful as a monitor for long-unattended periods, the stability of the oscillator frequency is critical. This requirement adds a great deal of sophistication to the instrument. The cost for a stabilized spectrometer would be about twice that of the instrument described in the preceding paragraph. Of course, it should be understood that much of the electronics needed for scanning (or stabilization) and signal processing can be integrated, resulting in a considerable reduction of size and cost. Since such technology is already well developed it would likely be used in any commercialization. This example is given as an estimate of the actual cost for construction of a single-gas monitor from commercially available parts. Other simple designs that are more amenable to commercialization are given in the following paragraphs.

The simplest spectrometer design is one in which the Gunn effect diode is used as both an oscillator and detector. Such an instrument is shown in block form in Fig. 5-13. A Gunn effect diode and a tuning diode (a varactor or voltage-variable capacitor) are located strategically within the waveguide. The waveguide is shorted in this case so that it becomes resonant for certain wavelengths (calculated to be resonant near the absorption frequency of the gas to be monitored). A vacuum window separates the diodes from the gas in the absorption cell. By applying a tuning voltage to the varactor diode, the resonant frequency of the entire cavity, including the absorption cell, changes over a range of several tens of megahertz. Since the Gunn effect diode oscillators at the resonant frequency of the cavity also, the microwave radiation is simultaneously tuned with the resonant frequency of the cavity. The gas to be monitored can be sampled continuously by pumping it through a gas-permeable membrane [165]. The tuning voltage is modulated by a sinusoidal signal at 250 Hz.

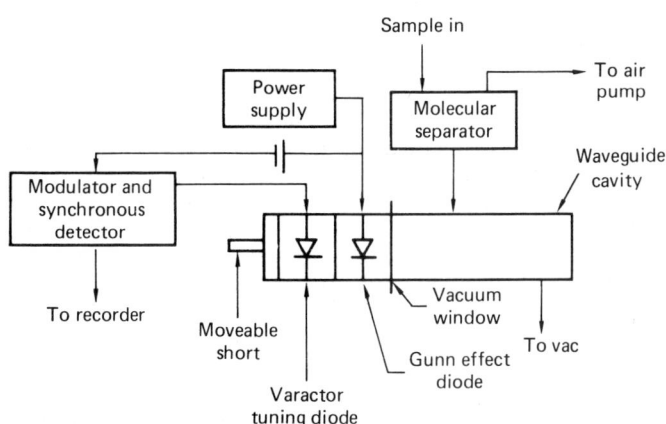

Fig. 5-13. Very simple MRS design in which Gunn effect diode is both oscillator and detector.

When a gas resonance is excited within the cavity, a signal at 250-Hz frequency appears on the bias lead to the Gunn effect diode. This signal can be synchronously detected with the 250-Hz reference to obtain a first derivative signal of the absorption as the frequency is tuned over the line. The output is fed back to the tuning diode (the varactor) in such a way as to stabilize the oscillator at exact coincidence with the peak center of the gas absorption. Subsequently, another synchronous detection is made at twice the 250-Hz frequency of the signal appearing at the bias lead, and this provides a d.c. output that is proportional to the amount of gas absorbing the microwaves. This type of a simple spectrometer has been used to detect water-vapor content in pure gases at the UCLLL with sensitivity to about 1 p.p.m. [166] (this sensitivity is only attainable with a membrane separator). The spectrometer described here is simple and inexpensive; it could probably be manufactured at a cost under $1000. Its reliability and ruggedness have been proven for laboratory use at the UCLLL over the past 3 years.

Another simple spectrometer has been designed at the UCLLL that also makes use of the Gunn effect diode as a sensitive detector. It is shown schematically in Fig. 5-14. This spectrometer is operated like a radiofrequency-balanced bridge of the type used for capacitive reactance measurements; in this case the gas absorption causes unbalance and is thus detected. The Gunn effect diode has been previously used as a "heterodyne" detector by Lazarus, Novak, et al. [167], where it is shown to be very sensitive in this mode because the diode acts as its own local oscillator and has a conversion gain rather than a loss. The "heterodyne" principle is used in the spectrometer shown in Fig. 5-14 in the following manner: Oscillator No. 1 is chosen to have a center frequency coinciding with the gas absorption to be monitored. Oscillator No. 2 has a frequency of 30 MHz above (or below) that of oscillator No. 1. The radiation from each oscillator passes through the waveguide-absorption cell to the other oscillator. The beat frequency (i.e., the difference between the fundamental frequencies of each oscillator), in this case 30 MHz, appears at the bias lead of each oscillator. This 30-MHz signal from each

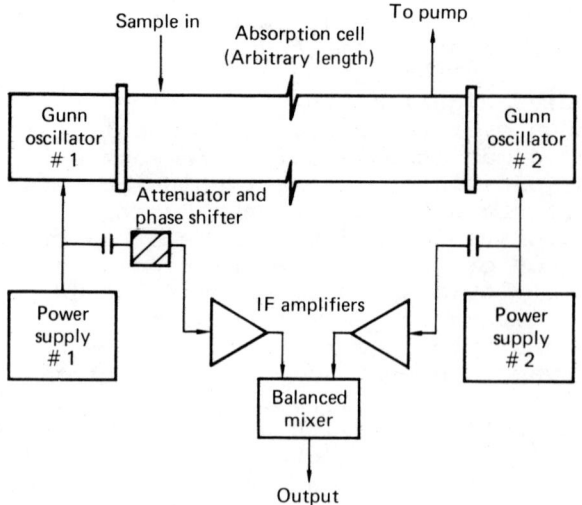

Fig. 5-14. Simple microwave spectrometer using Gunn effect diodes as heterodyne detectors.

detector is amplified and then input to a balanced mixer. The output of the mixer is nulled by adjusting the amplitude and phase of one input with respect to the other. When a gas absorbs some of the radiation from the oscillator that is tuned to its resonant peak, the signal at the other oscillator changes (in both amplitude and phase), thus unbalancing the balanced mixer. The phase shift causes a d.c. signal that is proportional to the gas-absorption strength, and it can be calibrated to represent the concentration of the absorbing gas.

This spectrometer is also simple and very sensitive; 2 p.p.m. of ammonia has been detected with a prototype. This system would be the most amenable to mass production and economical as well (<$1,000). One major drawback to this design is that it is extremely sensitive to temperature changes and is thus (in this simple form) somewhat limited for long-term monitoring applications. However, with just a bit more sophistication, this problem could be entirely eliminated.

Other simple designs have been proposed; some have even been constructed and proven operational. None, however, has had rugged field tests of the type necessary to qualify it as a reliable instrument for critical applications. It is hoped that in the near future instrumentation will be developed for a specific application in industry in which the technique can demonstrate its real capabilities as a monitor.

5.8 PROCESS-CONTROL SENSORS

There is a growing interest, particularly in chemical-, oil-, and energy-related industries, for compound specific process-control monitors. Bulk-effect sensors are still the standard with these industries, but selective detectors offer much more control for product optimization in many processes. Very few process-control sensors are compound selective, quantitative, and have real-time output capability. Microwave rotational spectroscopy has all of these properties plus compactibility, simplicity, and reliability.

Although it has great potential for process control applications, not much work has been done with MRS. However, several industries have shown interest in the past few years, and some work has been done recently by a group at the NBS in Washington, D.C.

One of the applications looked at by that group for which MRS is particularly well suited, is monitoring HCN gas during its synthesis. One reaction that produces HCN is ammonia with methane over a platinum-alloy screen catalyst at 1000°C,

$$CH_4 + NH_3 + 1.5O_2 \rightarrow HCN + 3H_2O.$$

Optimum yields of HCN are obtained when less oxygen (air) is used than indicated in this above equation to prevent excessive oxidation of ammonia to nitrogen oxides and methane to carbon oxides. Microwave rotational spectroscopy can be used to monitor NO, NO_2, or N_2O, whichever species are most indicative of the oxidation, as well as monitoring NH_3 and HCN. Thus the optimum concentration of O_2 could be automatically controlled by continuous differential relative intensity comparison (and feedback signals from the spectrometer). Also, in some HCN-production processes, HCN decomposition occurs since it is formed at temperatures well above the decomposition level. The carbon deposit on the catalyst screen then reduces its activity. Microwave rotational spectroscopy can be used to monitor NH_3 versus HCN to derive a signal that could be used to control the temperature of the reaction more precisely.

Another example where MRS could be useful as a process monitor is in methanol synthesis [168]. Methanol (CH_3OH) can be produced in the reversible reaction of carbon monixide (CO) and hydrogen at temperatures (100 to 400°C) and pressures up to 1000 atmospheres,

$$CO + 2H_2 \rightarrow CH_3OH.$$

A number of side reactions occur; for example,

$$CO + 3H_2 \rightarrow CH_4 + H_2O$$

$$2CO + 2H_2 \rightarrow CH_4 + CO_2$$

$$2CO \rightarrow C + CO_2$$

$$xCO + yH_2 \rightarrow \text{high-molecular-weight alcohols (etc.)}$$

$$CH_3OH \rightarrow H_2CO + H_2.$$

All but the last of these undesired reactions consume CO, and some poison the catalysts. The transitions in Table 5-7 show that many of the species involved in these reactions can be detected by MRS. It is possible to monitor each of these species separately or by differential comparison and thus allow an on-line control of the overall reaction.

TABLE 5-7. TYPICAL ABSORPTION COEFFICIENTS OF COMMON REACTION CONSTITUENTS[a]

Molecule	Transition Upper State Lower State[b]	Transition Frequency (MHz)	Absorption Coefficient (cm^{-1})
OCS	$2 \leftarrow 1$	24,325.92	5×10^{-5}
HCN	$9 \leftarrow 9, \nu_2 = 1$	20,181.40	2.6×10^{-6}
	$10 \leftarrow 10, \nu_2 = 1$	24,660.31	3.1×10^{-6}
CH_3CN	$1 \leftarrow 0$	18,397.70	1×10^{-5}
	$2 \leftarrow 1$	36,795.40	1×10^{-4}
NH_3	$6_1 \leftarrow 6_1$	18,391.60	2×10^{-6}
	$5_2 \leftarrow 5_2$	20,371.48	2×10^{-5}
	$5_5 \leftarrow 5_5$	24,532.94	4×10^{-4}
H_2O	$6_{1,6} \leftarrow 5_{2,3}$	22,235.08	1×10^{-5}
H_2CO	$9_{2,7} \leftarrow 9_{2,8}$	22,965.64	5×10^{-6}
CH_3CHO	$13_{2,11} \leftarrow 13_{2,12}A$	22,208.57	3×10^{-7}
	$6_{1,5} \leftarrow 6_{1,6}A$	22,346.47	3×10^{-7}
CH_3OH	$9_2 \leftarrow 10_1 A+$	23,121.20	2×10^{-6}
	$7_1 \leftarrow 7_1 A$	23,347.53	5×10^{-7}
	$10_1 \leftarrow 9_2 A-$	23,444.82	4×10^{-6}
trans-CH_3CH_2OH	$17_{7,10} \leftarrow 18_{6,13}$	22,531.90	2×10^{-7}
	$9_{2,8} \leftarrow 8_{3,5}$	23,130.20	1.5×10^{-7}
	$14_{6,5} \leftarrow 15_{5,11}$	23,177.55	2×10^{-7}
$(CH_3)_2CHOH$[b]		22,231.52	1.5×10^{-7}
		22,588.05	1.5×10^{-7}
		23,108.88	2×10^{-7}

[a]Source: Lovas [168].

[b]Assignment of transition uncertain; measurement from W. F. White (private communication).

Some work has been done to characterize the microwave spectrometer as a process control sensor, and experiments have been carried out to demonstrate its operation in a closed loop [169]. Figure 5-15 shows the simplified model describing the experiments to test the frequency response of a microwave spectrometer system. Figure 5-16 represents the actual setup of the experiment with the microwave spectrometer as the compound selective detector. Air is bubbled through methanol to prepare a constant concentration of methanol vapor in air. The mixture is flowed over the surface of a dimethyl silicone membrane that acts as an interface between atmosphere and the low pressure in the microwave absorption cell. The microwave spectrometer itself is tuned to detect only the methanol in the mixture, and its output is used to control a valve that regulates the flow of diluent air in the mixture. Figure 5-17 defines the transfer function of such a system assuming it to be a linear one. For a system disturbance, that is for $\Delta A_1(S)$, the overall transfer function is given by:

$$\frac{A_0(S)}{A_1(S)} = \frac{1}{e^{\tau_2 S}(1 + \tau_1 S) + G_c H_m (A_2/F_1)} \quad (5-1)$$

where

τ_1 = volume of cell/flow rate out of the bubbler;
τ_2 = volume of cell/flow rate out of control valve;
A_2 = concentration of diluent;
A_1 = concentration of methanol;
G_c = gain of controller; and
H_m = gain of microwave instrument.

For a change in set point:

$$\frac{A_0(S)}{A_{0d}(S)} = \frac{G_c(A_2/F_1)}{e^{\tau_2 S}(1 + \tau_1 S) + G_c H_m (A_2/F_1)} \quad (5-2)$$

where A_0 is the output and A_{0d} is the setpoint.

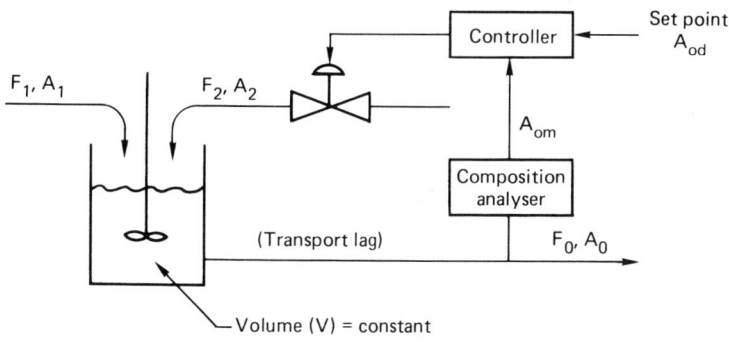

Fig. 5-15. Microwave spectrometry for process control: the model.

By measuring the actual system response to system disturbances, Eqs. (5-1) and (5-2) can be used to determine which elements in the system dominate its response time. For example, these equations can be used to determine the gain required for a given set of other conditions so that the system response will be limited only by the transducer response itself. Figure 5-18 shows the system for open-loop frequency-response measurements and time-domain data taken to measure the microwave system response to a sample impulse. Methanol vapor is momentarily admitted to the input, and the output of the spectrometer is then monitored as a function of time. These data are digitized and then transformed into the frequency domain to determine the effective time constant of the system. For the setup shown in Fig. 5-18 the frequency response was 0.7 Hz. It can be shown that this slow response is due mainly to membrane permeability in such a system [170]; the response of the microwave spectrometer alone is typically of the order of milliseconds.

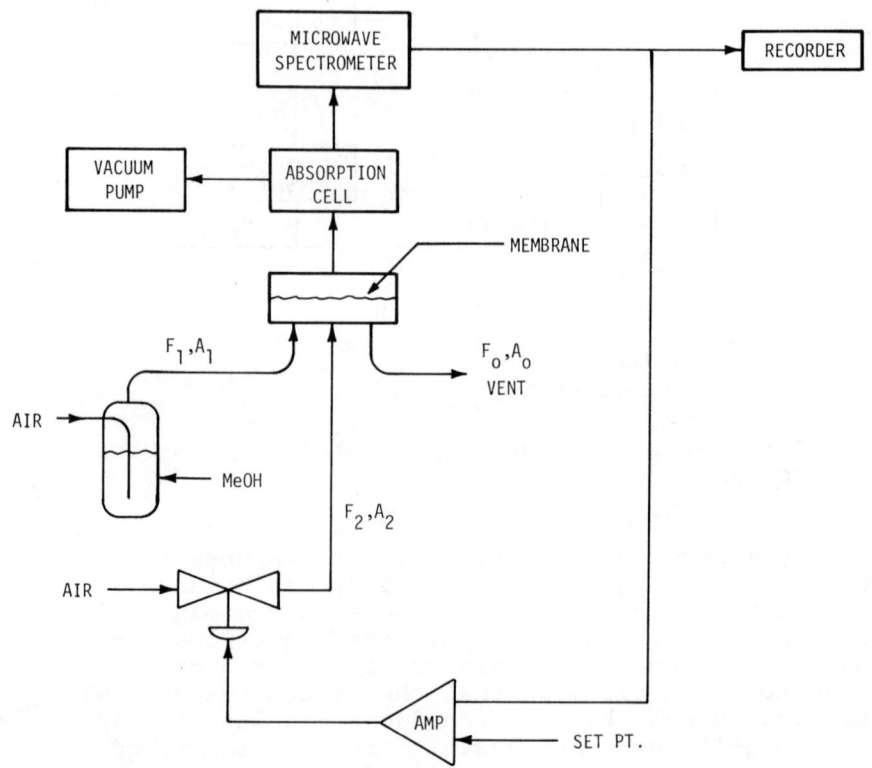

Fig. 5-16. Schematic of experiment to measure frequency response of microwave spectrometer with membrane interface.

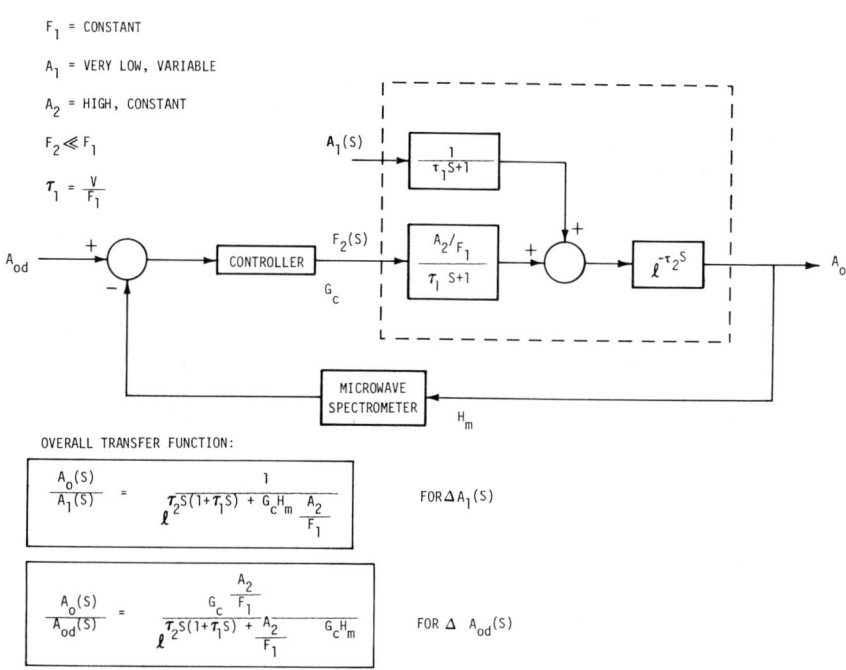

Fig. 5-17. Transfer function of linear system with MRS.

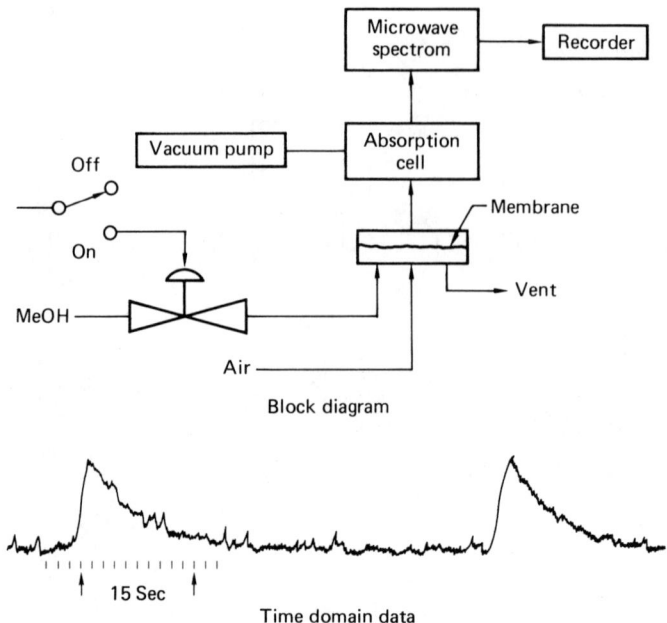

Fig. 5-18. System used for open-loop frequency-response measurements and time-domain data from sample impulse.

Figure 5-19 shows how the system responds to a change in set point and to a system disturbance. Note that the system is underamped, that is, it does not smoothly seek the final value.

An actual test of the operation of a microwave spectrometer in closed-loop operation is shown schematically in Fig. 5-20. In this experiment the objective is to show that the liquid fraction of methanol produced in a distillation of a 50:50 molar mixture of methanol and ethanol can be controlled by the microwave spectrometer. The vapor above the mixture is continuously sampled for methanol by the microwave spectrometer; its output controls the current to the heater of the distillation flask. For each set point the distillation is run for 2 h, and then the

CLOSED LOOP DATA

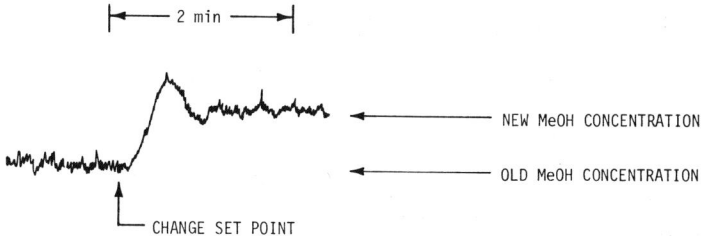

RESPONSE TO A CHANGE IN SET POINT

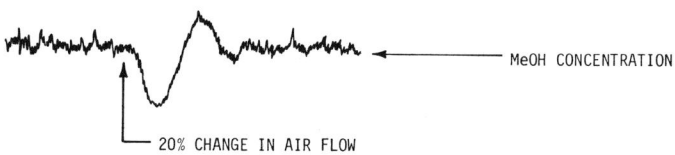

RESPONSE TO A SYSTEM DISTURBANCE

Fig. 5-19. System response to change in set point and to system disturbance (i.e., simulates a concentration change).

condensed liquid is analyzed by gas chromatography to determine the liquid fraction of methanol. Data are plotted on the MeOH/EtOH equilibrium diagram in Fig. 5-21 for several control settings. The response is linear within the operating regime. This work clearly shows the utility of compound-selective detection for feedback-controlled operation. The microwave spectrometer can clearly be used to control processes, even where there is a rather complex mixture of compounds to contend with.

If inexpensive, reliable monitors can be developed for specific processes, MRS is likely to become very popular for process-control applications. There is a

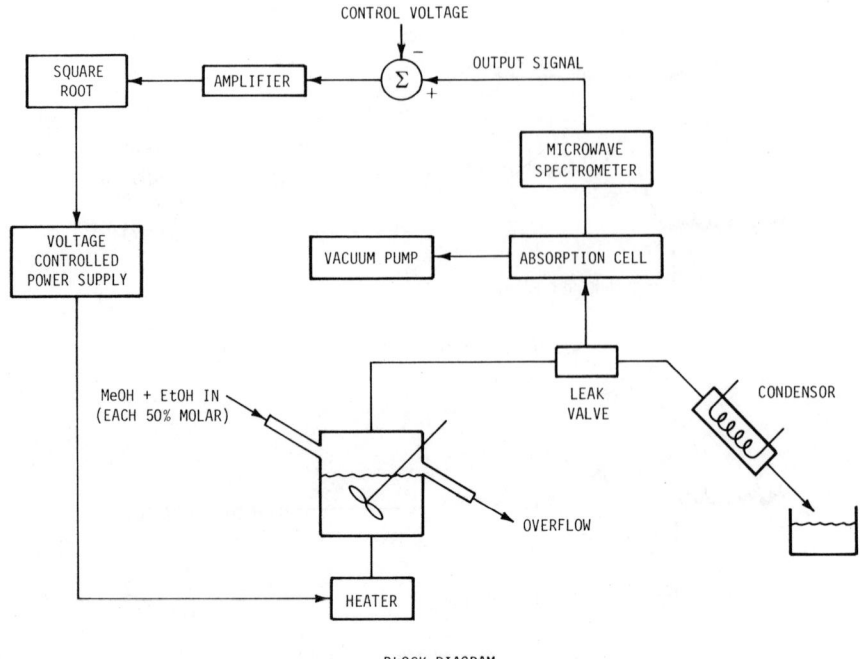

Fig. 5-20. Process-control simulation experiment with MRS. Microwave spectrometery controls distillation fraction of MeOH in MeOH-EtOH mixture.

growing interest for such devices by the petrochemical industry, whereas energy-related processes in general are in need of advanced instrumentation.

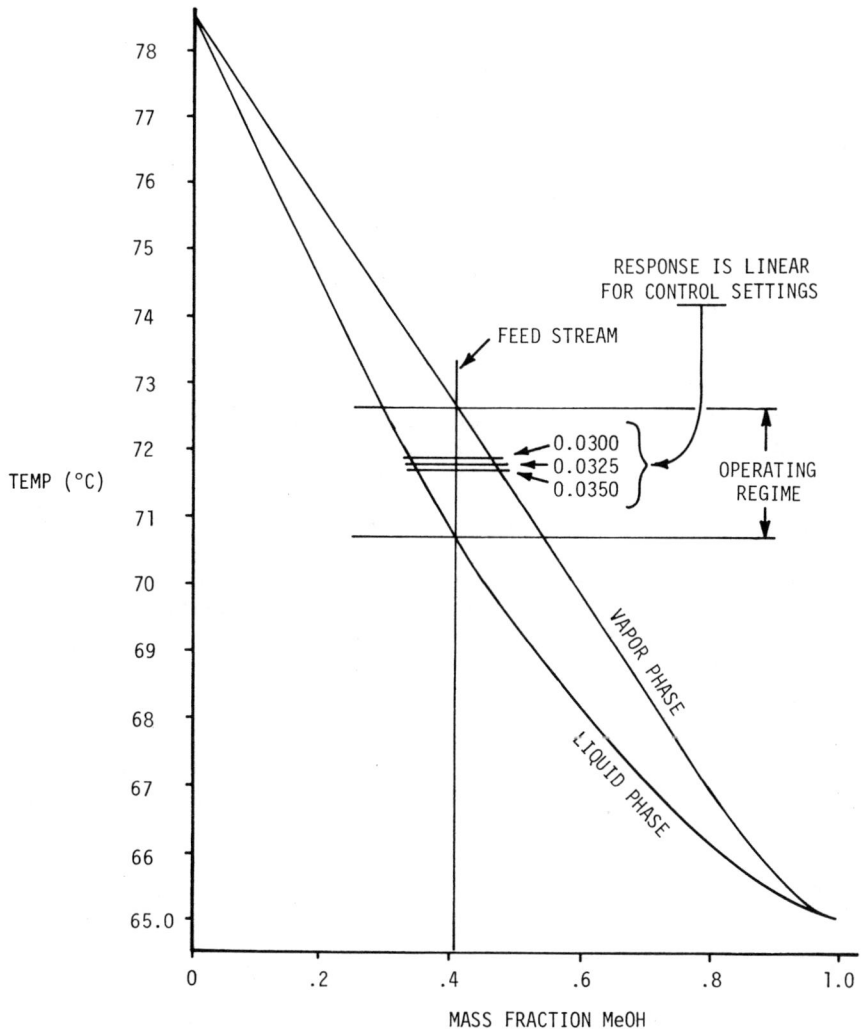

Fig. 5-21. Diagram of MeOH-EtOH equilibrium showing operating region of experiments and linearity of system response (1 atm).

5.9 PORTABLE GAS MONITORS

With the advent of solid-state microwave sources in the late 1960s, the concept of portable, field-usable microwave spectrometers became a reality. Prior to that time the microwave oscillators were bulky and required high voltages, and no attempts were made to develop portable instrumentation.

In 1970 one of the authors developed a very simple spectrometer (discussed in Section 3-1) with a Gunn effect diode and a resonant-cavity absorption cell. This led the way to compact, portable MRS instruments that were subsequently developed for air-pollution and occupational air-quality measurements. These instruments are described in this section. The concept of a truly portable, battery-operated, hand-held "sniffer" is also discussed.

The first of the compact, special-purpose MRS spectrometers developed at the UCLLL was designed to detect only formaldehyde vapor in air [146]. The instrument was developed under contract for the Environmental Protection Agency in Research Triangle Park, North Carolina.

A block diagram of the instrument is shown in Fig. 5-22, and a photograph is shown in Fig. 5-23. The instrument, though compact, is quite complex. Most of the complexity results from the stiff specifications set by EPA: (1) long-term stability, (2) automatic (unattended) operation for extended periods, (3) ultra-sensitivity (0.03 p.p.m.), and (4) rapid response (90% full scale within seconds). A Gunn effect oscillator is the source, and a resonant cavity is the absorption cell.

The oscillator frequency and the cavity-resonant frequency are both stabilized by servoelectronics to ensure long-term stability. An oven-controlled, low-frequency oscillator is used as a reference to "synthesize" the microwave frequency through a series of multiplier and phase-lock loops. The resonant frequency of the cavity absorption cell is then stabilized to coincide with the oscillator frequency by comparing the phase of the microwaves before and after they pass through the cavity. The phase difference detected in this way is exactly zero when the cavity resonance matches the microwave frequency and has opposite polarity depending on the direction of mismatch. This "discriminant" type of signal is used to

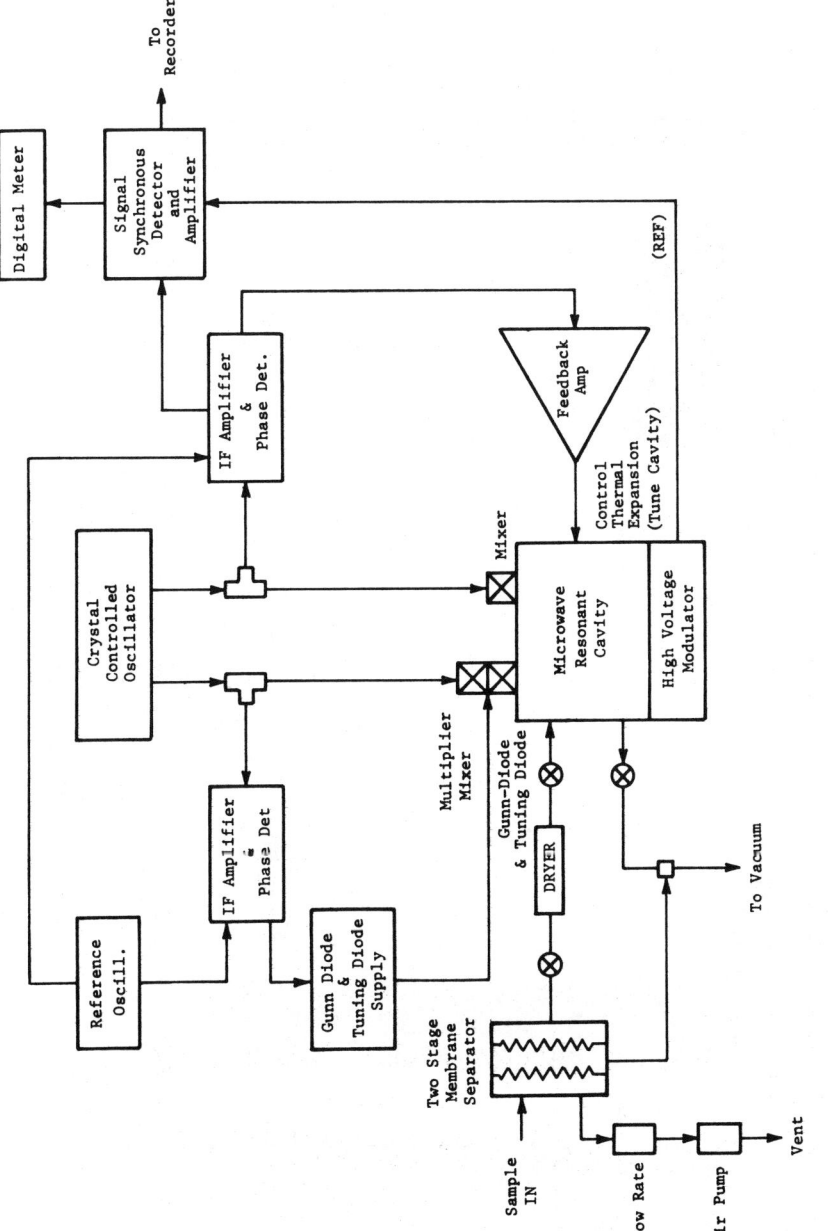

Fig. 5-22. Block diagram of portable microwave instrument to monitor formaldehyde vapor.

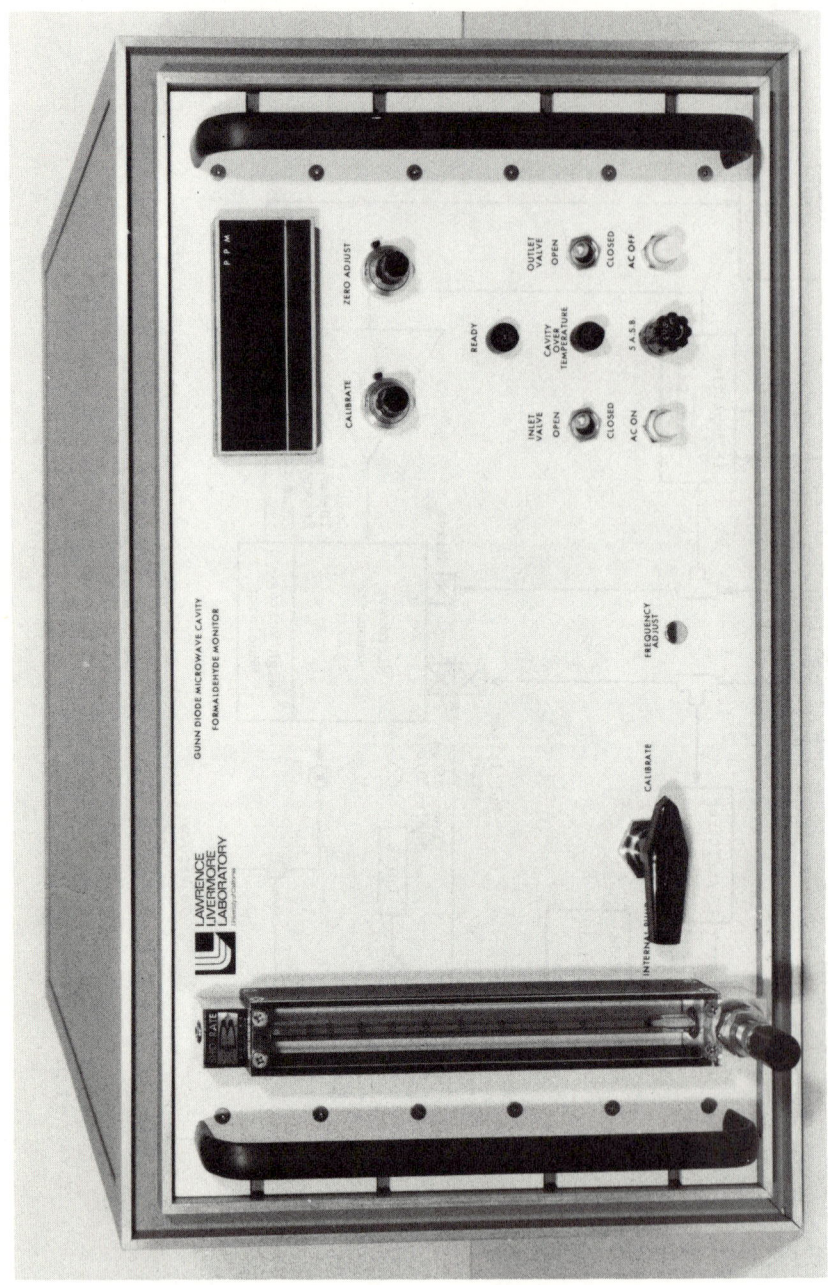

Fig. 5-23. Photograph of Gunn diode microwave formaldehyde monitor developed for EPA by the UCLLL.

control the mechanical spacing between the resonator reflectors to a very high degree of stability. Thus any environmental changes that would affect the synchronization of the source frequency and resonant frequency are compensated.

The sampling problems for formaldehyde were difficult to resolve in this instrument. A dimethyl silicone membrane was used as an interface to the atmospheric pressure from the low pressure in the absorption cell. Gases were sampled by continuous flow over the atmospheric side of the membrane; permeated gases were then pumped through the absorption cell. Early in the development of this instrument, however, two problems arose: (1) trapping of formaldehyde monomer on cool surfaces and (2) effect of water vapor on the permeation characteristics of the membrane for formaldehyde. Problem (2) was solved by using a water trap of P_2O_5 that was found to only slightly affect the formaldehyde throughput. Problem (1) was not satisfactorily solved and led to an extremely long response time for the instrument.

Figure 5-24 shows the response of the monitor to approximately 0.03 p.p.m. of formaldehyde in air. Real-time changes in formaldehyde concentrations as small as 0.03 p.p.m. could be detected with this instrument even with the problems of surface effects.

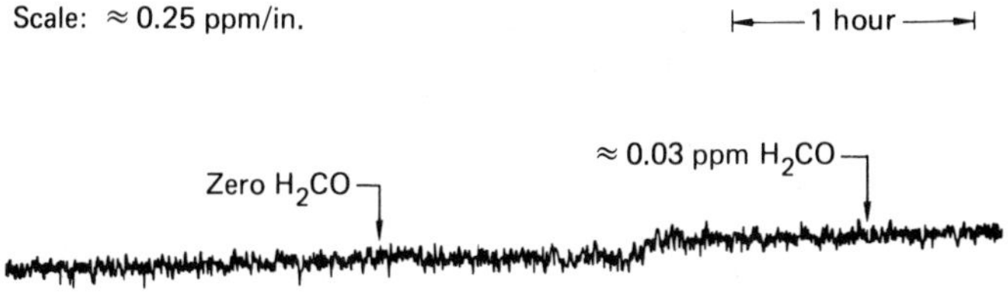

Scale: ≈ 0.25 ppm/in. |⎯⎯ 1 hour ⎯⎯|

Zero H_2CO ≈ 0.03 ppm H_2CO

Fig. 5-24. Minimum detectability of formaldehyde monitor. Shows response of instrument to 0.03 p.p.m. formaldehyde in air.

The calibration procedure used teflon permeation tubes filled with paraformaldehyde. At a constant temperature, the tube emitted a constant quantity of monomer vapor. The volume percent (or parts per million) could then be controlled by the flow of a carrier gas (N_2 or air) over the permeation tube. The resulting standard mixture was flowed over the atmospheric side of the membrane. Figure 5-25 shows the typical calibration curve for the instrument and indicates a comparison with the chromotropic acid test for formaldehyde. The apparent discrepancy between the measured values was ascribed to the loss of formaldehyde to the inlet lines and P_2O_5 trap in the microwave spectrometer.

Although the completed instrument as delivered to the EPA was capable of selective detection of small amounts of formaldehyde, it suffered major limitations because of the sampling problems and inadequate sensitivity for ambient formaldehyde measurements (<30 p.p.b. sensitivity was needed). Commercialization of this design did not result because of the lack of a market for formaldehyde-specific detectors. This situation may change, however, with the current emphasis on alternate fuels for internal combustion.

A similar design was used to develop an ammonia monitor for the California Air Resources Board [171]. The major difference between the ammonia monitor and the formaldehyde monitor was that the resonant cavity of the ammonia monitor, including all the inlet tubing and membrane separator, was placed in an oven. The sampling problems associated with adsorption of gases on the inlet surfaces were thus eliminated.

The ammonia instrument has been proven very sensitive and reliable. In laboratory tests it can detect a standard of 10 p.p.b. of ammonia in air. This represents the highest sensitivity ever achieved for the continuous detection of any gas by microwave spectrometry and compares well with any other technique currently used for ammonia detection in air.

To extend the sensitivity of MRS for almost any gas, a method of bulk preconcentration can be used. One technique that has worked well for ammonia preconcentration is the use of a chromatographic material to trap the ammonia gas from a flowing air stream, then after an appropriate trap time, heating it to drive off the trapped gas as one plug. This technique has also been used successfully by White at NASA [148] in unpublished work relating to the detection of

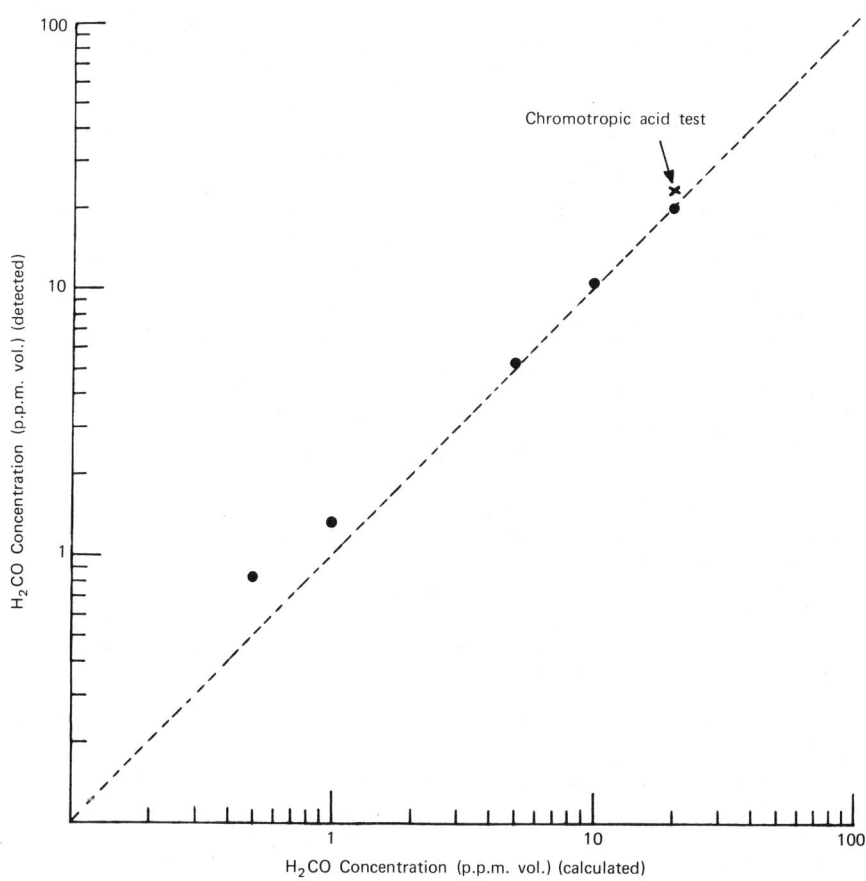

Fig. 5-25. Calibration curve for formaldehyde monitor. Comparison is shown with wet-chemical analysis of same standard.

fluorocarbons above sea water by MRS. This method is particularly appropriate to use with a microwave detector because MRS is sufficiently compound selective that other gases that may be trapped along with the species of interest (notably water vapor) do not interfere in the detection mode. This principle has worked so well for ammonia concentration that the sensitivity of MRS for detecting ammonia has been extended to sub-parts-per-billion quantities in air [147]. An instrument has been developed at UCLLL that automatically cycles the trapping and flashing process and then integrates the signal produced by the ammonia monitor. The preconcentrator/integrator and a version of the ammonia monitor that was developed for the California Air Resources Board is now field deployed in a mobile unit in the Imperial Valley of California for ambient ammonia monitoring. This instrumentation is proving to be a desirable alternative to the wet chemical techniques formerly used for these measurements.

The combination of the trap with the ammonia monitor is shown schematically in Fig. 5-26. A typical cycle of operation with this combination is illustrated in Fig. 5-27. The trace shown is the response of the microwave ammonia monitor whose gas input is supplied from the preconcentrator/integrater unit. The first part of the trace shows the output of the spectrometer to a continuous flow of about 4 p.p.m. of ammonia. Then the trap is switched into the line, and the instrument response drops to zero. After a long period of time (~ 30 min) the instrument starts to detect the ammonia that is leaking through the saturated trap. A very large signal is observed as the trap is heated and flushed with clean air. The function of the preconcentrator/integrator is to automatically cycle the trapping and heating functions and to integrate the sample peak, subtract it from the baseline, and then display the difference as the net signal of detected ammonia. The trapping time (which results in different minimum detectability for ammonia), is selectable from 10 to 50 min in 10-min intervals.

The concept of portable gas monitors can be directly extended to multiple gas detection in a single instrument. It is possible to package a special-purpose spectrometer of the conventional design (i.e., synthesized frequencies, a stable microwave oscillator, and a wave-guide Stark cell) into a fairly small package. A fully functional spectrometer can be made quite portable by using a solid-state broad-band YIG-tuned oscillator, spiraled waveguide with Stark septum included, a compact high-frequency synthesizer (available commercially), and microprocessor control. The vacuum pump required for the reduced pressure in the absorption cell limits overall

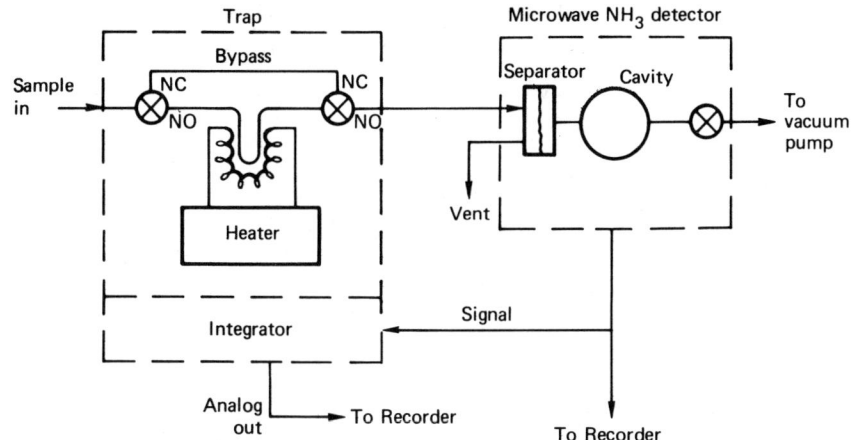

Fig. 5-26. Schematic of preconcentration/integrator in operation with ammonia detector.

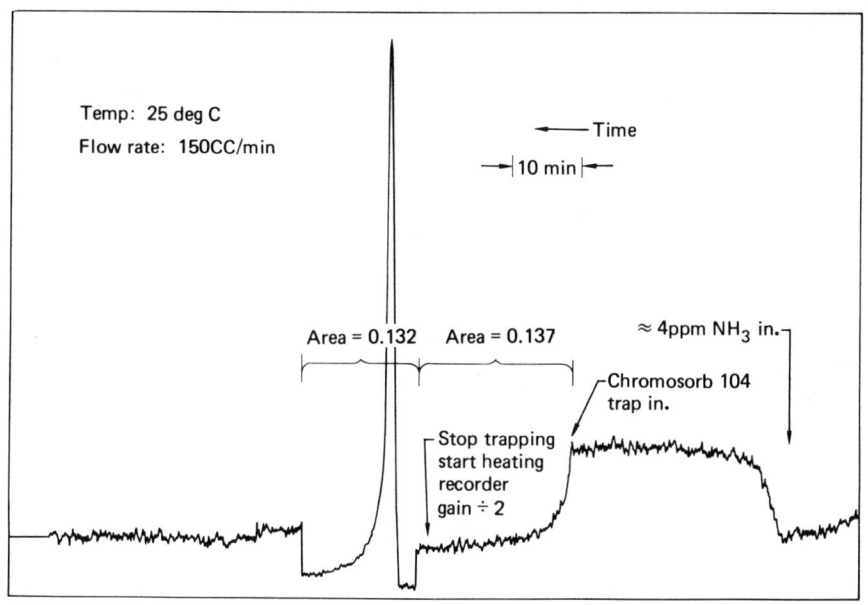

Fig. 5-27. Operation of preconcentrator/integrator for ammonia vapor as detected with microwave spectrometer

portability, but fairly small pumps are now available with sufficient capacity for a microwave rotational spectrometer.

A different concept for microwave source stabilization is being developed at the UCLLL for a portable multigas monitor to be used by the NIOSH [154]. The instrument must be sufficiently portable for a single person to carry into an occupational environment to test the air for potentially hazardous levels of ordinary solvents. Rather than synthesize the source frequency by multiplication and phase-lock loops, this portable instrument will use permeation tubes filled with the gas to be monitored, both to stabilize the source frequency and to calibrate the instrument for that gas.

The concept works as follows: The permeation tubes emit a constant amount of the gases to be monitored into a reference waveguide absorption cell. Here they are detected in the usual way by an oscillator tuned to resonance peak, but the oscillator is modulated in such a way that the gas absorption provides a "discriminant" signal that can be fed back to the source oscillator to stabilize it at the resonant peak of the reference gas. Some of the microwave power is coupled out of this circuit and passed through the analytical absorption cell where the analysis of the sampled air is accomplished. In addition to providing a reference gas to stabilize the oscillator, the same permeation devices can provide a constant reference gas for calibration purposes, all within the same small instrument. Some advantages of this approach are: (1) simplicity of electronics needed for frequency lock, (2) frequency stabilized to an "absolute" reference (i.e., gas absorption) and never requires fine tuning, (3) calibration standards provided within the instrument, and (4) new species monitored simply by adding the proper permeation tube to the instrument.

The instrument currently being developed for NIOSH will detect the gases and vapors listed in Table 5-8. It will have a push-button selector for each of the gases to be monitored. Additional gases can be added at a later date by adding the appropriate permeation tubes (which can be obtained commercially). The entire instrument, excluding the vacuum pump, will weigh less than 23 Kgm and will operate either from a battery or from a.c. power. The instrument will be delivered to the NIOSH by the summer of 1978.

TABLE 5-8. THRESHOLD LIMIT VALUES AND CALCULATED MINIMUM DETECTION LIMITS BY MRS

Compound	NIOSH Limit[a] (p.p.m.)	Minimum Detection Limit
Acetonitrile	40	8
Acetaldehyde	100	20
Acetone	1000	200
Carbonyl sulfide	60	12
Ethanol	1000	200
Ethylene oxide	50	10
Isopropyl alcohol	400	80
Methanol	200	40
Propylene oxide	100	20
Sulfur dioxide	5	1

[a]Taken from Threshold Limit Values for Chemical Substances and Physical Agents in the Workroom Environment with Intended Changes for 1973. With permission from the Secretary-Treasurer, American Conference of Governmental Industrial Hygienists, P. O. Box 1937, Cincinnati, Ohio 45201.

The ultimate in portability for MRS, namely, a hand-held "sniffer" type of instrument, is feasible and could be developed in the near future. The major problem to overcome in such an instrument is the vacuum requirement. Very small vacuum pumps do not currently exist, although they appear to be feasible. Other concepts for pumping out the absorption cell between sample injections are being considered.

One of the most promising ideas is to use a combination of sorption material and getter pellets to pump the absorption cell after a sample is analyzed. The absorption cell would be rough-pumped once, prior to field operation, or after replacement of the sorption and getter material. Then the instrument would be usable for analyzing up to 50 samples, with each leaked into the absorption cell, analyzed, and then pumped out.

It is believed that a battery operated Gunn effect diode and a crystal detector could be placed within a wave guide that was machined into a metal plate. The length of the absorption cell could be quite long (1-m length is reasonable) if machined as a spiral in a 3" × 6" × 1" block of metal. A microprocessor (battery operated) could control the sweep functions and handle the signal information. Thus it is conceivable that a microwave spectrometer could be reduced to the size of a hand-held calculator with enough sensitivity to be useful for identifying hazardous levels of gases in air. There appears to be sufficient interest by various governmental agencies for a device of this potential, and it is likely to be developed within the next couple of years.

Microwave rotational spectrometers can be compact, low-power, and portable as shown by the preceding examples. They need not be sophisticated. Since they offer, in general, a much higher degree of selectivity for gas detection than IR or other nonspectroscopic techniques and have other favorable characteristics for gas monitors, they could easily replace other techniques for portable applications. What seems to be lacking is a clear demonstration of their utility as compared with currently used techniques and an indication that they can be manufactured at competitive cost. The authors feel that MRS can be competitive with other portable instruments and that these spectrometers will find popularity in the future.

5.10 COMBINED GAS CHROMATOGRAPHY/MICROWAVE ROTATIONAL SPECTROMETERY

Within the past 7 years or so, extremely powerful analytical tools have been found by combining the separating characteristics of gas chromatographs with other sensitive (and sometimes selective) detectors. The most popular of these is the gas chromatograph/mass spectrometer combination, which is already well established and commercially available. The combination of a microwave rotational spectrometer with a gas chromatograph also has some outstanding characteristics, especially for quantitative gas analysis. In this section we describe some of the work that has been done with gas chromatography/MRS to demonstrate its potential analytical power.

Some of the principal advantages to combining a gas chromatograph and a microwave rotational spectrometer are:

1. There is virtually a 100% probability of component identification using just a single microwave absorption line.
2. The gas chromatograph does not need to provide complete component separation for an absolute component identification.
3. Accurate quantitative analysis can be made on components in an unresolved mixture.
4. The microwave absorption cell can be operated in a "sample and hold" mode that enables the operator to scan the microwave spectrum slowly if desired for maximum resolution and sensitivity.
5. The technique is nondestructive for the sample; there is no molecular breakdown or chemical reaction due to hot filament, electron bombardment, and so on.

It is seen here that this combination possesses some of the favorable advantages of both gas chromatograph/mass spectrometers and gas-chromatograph/IR detectors and, as such, is potentially capable of the same power of analysis as these two established techniques. For example, there have been some cases at the UCLLL for analysis of a complex mixtures of samples that were not to be destroyed by the analyses. Gas chromatograph/mass spectrometry was not applicable, the gas chromatograph/IR technique was not quantitative. The mixtures could be totally analyzed by the microwave spectrometer alone, but with a much lower overall sensitivity than is possible with the gas chromatograph, providing at least gross separation of the mixture components.

Some work has been done by Morrison [172] demonstrating the ease with which quantitative measurements can be accomplished with a gas chromatograph/MRS combination. One analysis problem involved a determination of the amount of methanol in a methanol-water-gasoline mixture that was used in alternate fuel research at the UCLLL. The analysis by gas chromatography alone was complicated by a partially unresolved peak for methanol, even after temperature programming the column. The eluted gases from the gas chromatography were sampled directly by the microwave spectrometer through a membrane separator of the type described earlier in this chapter. The MRS spectrometer was tuned to detect only methanol and clearly resolved the methanol peak from the rest of the mixture. Next, Morrison calibrated the combined gas chromatograph/MRS by injecting known quantities of methanol prior to injecting mixtures to be analyzed; the output of the microwave spectrometer was recorded for each injection. Figure 5-28 shows the traces for a typical analysis. With this technique, Morrison was able to quickly quantify the amount of methanol in the mixture to within 5% accuracy, which was more than sufficient for this particular case.

In another example [172] Morrison used a similar approach to determine the quantity of carbonyl sulfide in fractions from several oil-shale retort burns. The purpose of this work was to use several independent methods for comparison (a mass spectral analysis, gas chromatography with sulfur-specific detectors, and a gas chromatograph/MRS combination) to determine the sulfur-bearing species from the different shale types used. The calibration technique used by Morrison was similar to that described in the preceding paragraph. Known quantities of carbonyl sulfide at the 1 to 20-p.p.m. levels in nitrogen gas were prepared using a permeation device in a well-controlled oven. Samples were removed from the air-flow stream (after removal from the oven) by drawing out a sample in a syringe and injecting it into the gas chromatograph. The eluted peaks for standards as detected by the microwave spectrometer can be seen in Fig. 5-29 to provide a fairly linear calibration curve. Subsequently, the actual retort samples were analyzed and the results compared with the other methods. Table 5-9 shows the results of a typical analysis by two different methods, MRS and mass spectroscopy.

The ease with which the analysis can be made with the MRS has made it the preferred choice by oil-shale researchers at the UCLLL. Gas chromatography combined with MRS has some strong properties for complex mixture analyses and is expected to find use in future energy-related analyses like those just mentioned.

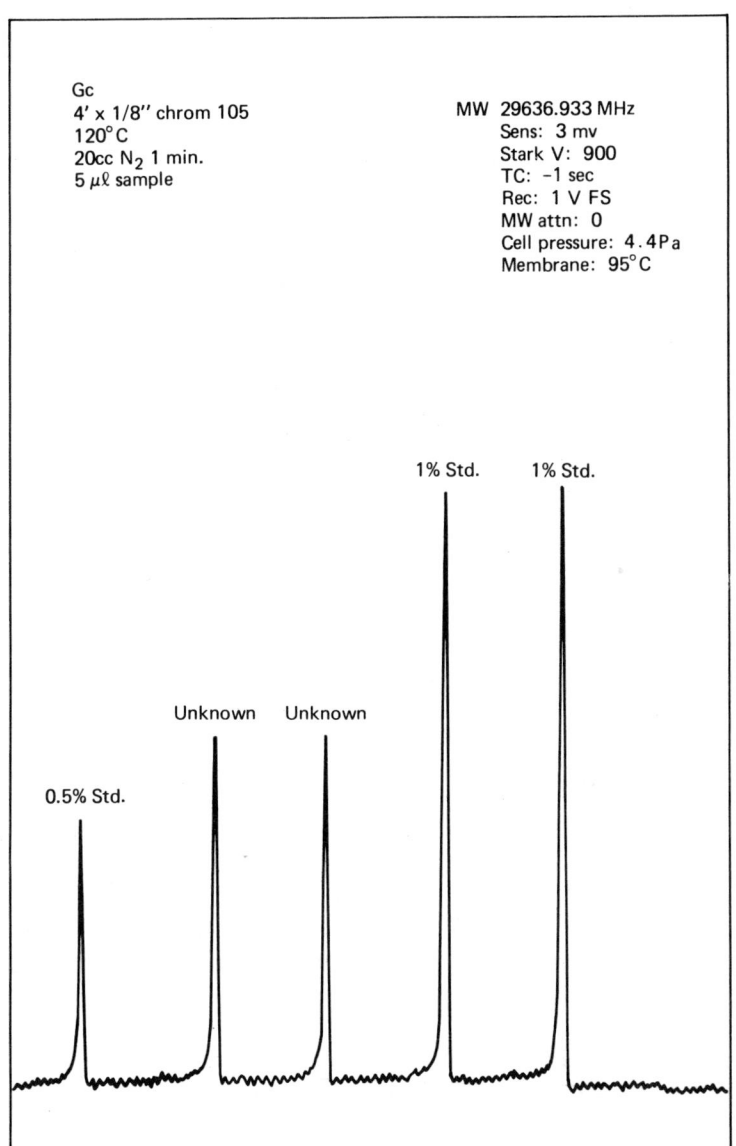

Fig. 5-28. Analysis of methanol-water-gasoline mixture by a combined gas chromatography/microwave spectrometer.

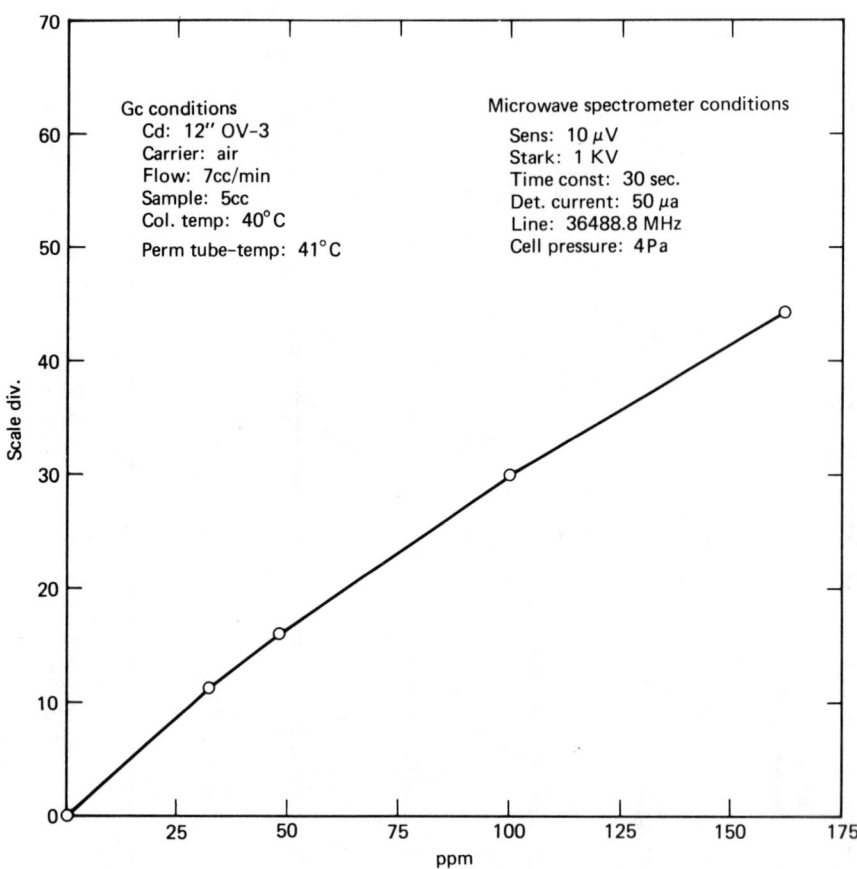

Fig. 5-29. Calibration for parts-per-million amounts of carbonyl sulfide in combination gas chromatography/microwave spectrometer.

TABLE 5-9. RESULTS OF AN ANALYSIS FOR COS AND SO_2 IN OIL-SHALE RETORT RESIDUAL GAS BY ALTERNATE METHODS

Time into run, h	Sample No.	COS, p.p.m.		SO_2, p.p.m.	
		Microwave	Mass-Spec.	Microwave	Mass
10.7	6	<2	—	<10	—
29.7	9	122	72	<10	—
41.7	11	82	65	12	—
66.7	17	72	66	<10	—

REFERENCES

1. B. Bleaney and R. P. Penrose, Nature, **157**, 399 (1946).

2. W. E. Good, Phys. Rev., **69**, 539 (1946).

3. C. H. Townes, Phys. Rev., **70**, 665 (1946).

4. W. Gordy and R. L. Cook, Microwave Molecular Spectra, Wiley, New York, 1970.

5. J. E. Wollrab, Rotational Spectra and Molecular Structure, Academic Press, New York, 1967.

6. C. H. Townes and A. L. Schawlow, Microwave Spectroscopy, McGraw-Hill, New York, 1955.

7. A. C. Legon and D. J. Millen, Molec. Spectrosc., **2**, 1 (1974).

8. D. R. Lide, Jr., Methods in Experimental Physics, **3**, Pt. A, 11 (1974).

9. J. Sheridan, MTP Int. Rev. Sci. Phys. Chem., Ser. 1, **12**, 251 (1973), Butterworths, London.

10. W. H. Flygare, Ann. Rev. Phys. Chem., **18**, 325 (1967).

11. Y. Morino and E. Hirota, Ann. Rev. Phys. Chem., **20**, 139 (1969).

12. H. D. Rudolph, Ann. Rev. Phys. Chem., **21**, 73 (1970).

13. E. B. Wilson, Science, **162**, 59 (1968).

REFERENCES

14. L. H. Scharpen and V.W. Laurie, Analytic Chemistry, Ann. Rev., $\underline{44}$, 378R (1972).

15. F. J. Lovas, Anal. Instr., $\underline{12}$, 103 (1974).

16. R. L. Hilderbrandt, J. Chem. Phys., $\underline{51}$, 1654 (1969).

17. H. B. Thompson, J. Chem. Phys., $\underline{37}$, 3407 (1967).

18. H. C. Allen and P. C. Cross, Molecular Vib-Rotors, Wiley, New York, 1963.

19. B. S. Ray, Z. Physik, $\underline{78}$, 74 (1932).

20. S. C. Wang, Phys. Rev., $\underline{34}$, 243 (1929).

21. R. A. Beaudet, Ph.D. dissertation, Harvard University, Cambridge, Mass. (1964).

22. F. Kneubuhl, T. Gaumann, and H. Gunthard, J. Molec. Spectrosc., $\underline{3}$, 349 (1959).

23. G. F. Pollnow and C. S. C. Chung, J. Chem. Ed., $\underline{50}$, 794 (1973).

24. A. B. Delfino and K. P. Ramaprasad, J. Molec. Struct., $\underline{25}$, 293 (1975).

25. J. K. G. Watson, J. Chem. Phys., $\underline{46}$, 1935 (1967); J. Chem. Phys., $\underline{48}$, 181 (1968).

26. D. Kivelson and E. B. Wilson, Jr., J. Chem. Phys., $\underline{20}$, 1575 (1952); J. Chem. Phys., $\underline{21}$, 1229 (1953).

27. W. H. Kirchhoff, J. Molec. Spectrosc., $\underline{41}$, 333 (1972).

28. H. G. B. Casimir, On the interactions between atomic nuclei and electrons, Teylers Tweede Genootschap, E. G. Bohn, Haarlem, The Netherlands, 1936.

29. K. B. McAfee, Jr., H. R. Hughes, and E. B. Wilson, Jr., Rev. Sci Instr., $\underline{20}$, 821 (1949).

REFERENCES

30. R. H. Johnson and M. W. P. Strandberg, Phys. Rev., 86, 811 (1952).

31. B. Karplus, Phys. Rev., 73, 1027 (1948).

32. H. A. Lorentz, Proc. Amsterdam Acad., 8, 591 (1906).

33. J. H. Van Vleck and V. F. Weisskopf, Rev. Mod. Phys., 17, 227 (1945).

34. D. K. Coles, Adv. Electronics, 2, 300 (1950).

35. Microwave Spectral Tables, Monograph 70 Series:
    Vol. 1, Diatomic Molecules;
    Vol. 3, Polyatomic Molecules with Internal Rotation;
    Vol. 4, Polyatomic Molecules without Internal Rotation;
    National Bureau of Standards, U. S. Department of Commerce, Washington, D.C.

36. W. F. White, Langley Research Center, Hampton, VA 23665, NASA Technical Notes:

    TND-8053, Microwave Spectral Line Listing;
    TND-7450, Microwave Spectra of Some Sulfur Compounds;
    TND-8002, Microwave Spectra of Some Chlorine and Fluorine Compounds;
    TND-7904, Microwave Spectra of Some Volatile Organic Compounds.

    W. F. Kolbe, B. Leskovar, and H. Buscher, Absorption Coefficients of Sulfur Dioxide Microwave Rotational Lines, UCLBL Report 3249, March 10, 1975.

    W. F. Kolbe, H. Buscher, and B. Leskovar, Microwave Absorption Coefficients of Atmospheric Pollutants and Constituents, UCLBL Report 4467, February 15, 1976.

37. Microwave spectra of molecules of astrophysical interest:

    a. Formaldehyde, formamide, and thioformaldehyde; D. R. Johnson, F. J. Lovas, and W. H. Kirchhoff, J. Phys. Chem. Ref. Data, 1 (4),

REFERENCES

1011 (1972).

b. Methylenimine: W. H. Kirchhoff, D. R. Johnson, and F. J. Lovas, J. Phys. Chem. Ref. Data, 2(1), 1 (1973).

c. Methanol: R. M. Lees, F. J. Lovas, W. H. Kirchhoff, and D. R. Johnson, J. Phys. Chem. Ref. Data, 2 (2), 205 (1973).

d. Hydrogen sulfide: P. Helminger, F. C. DeLucia, and W. H. Kirchhoff, J. Phys. Chem. Ref. Data 2(2), 215 (1973).

e. Water vapor: F. C. DeLucia, P. Helminger, and W. H. Kirchhoff, J. Phys. Chem. Ref. Data 3(1), 211 (1974).

f. Carbonyl sulfide and hydrogen cyanide: A. G. Maki, J. Phys. Chem. Ref. Data 3(1), 221 (1974).

g. Carbon monoxide, carbon monosulfide, and silicon monoxide: F. J. Lovas, D. H. Drupenie, J. Phys. Chem. Ref. Data 3(1), 245 (1974).

h. Sulfur monoxide: G. Tiemann, J. Phys. Chem. Ref. Data 3(1), 259 (1974).

i. Acetaldehyde: A. Bauder, F. J. Lovas, and D. R. Johnson, J. Phys. Chem. Ref. Data 5(1), 53 (1976).

j. Silicon sulfide: E. Tiemann, J. Phys. Chem. Ref. Data 5(4), 1147 (1976).

k. Iso cyanic acid: G. Winnewisser, W. H. Hocking and M. C. L. Gerry, J. Phys. Chem. Ref. Data 5(1), 79 (1976).

38. H. W. Harrington, J. Chem. Phys., $\underline{46}$, 3698 (1967).

39. H. W. Harrington, J. Chem. Phys., $\underline{49}$, 3023 (1968).

40. H. W. Harrington, J. R. Hearn, and R. F. Rauskolb, Hewlett Packard Journal, June (1971), pp. 2-12.

REFERENCES

41. S. Armstrong, Appl. Spectrosc., **23**, 575 (1969).

42. W. F. White, Automation and the Microwave Spectrometer, presented at the Ninth National Meeting of Society of Applied Spectroscopy, 1970.

43. L. H. Scharpen and R. F. Rasukolb, Anl. Chem., **44**, 2010 (1972).

44. J. Cuthbert, E. J. Denney, E. J. Silk, R. Stratford, J. Farren, T. L. Jones, D. Pooley, R. K. Webster, and R. H. Wells, J. Phys. E: Sci. Instr., **5**, 698 (1972).

45. G. Winnewisser, M. Winnewisser and B. P. Winnewisser, Millimeter Wave Spectroscopy, in MTP International Review of Science, Vol. 3, D. A. Ramsay, Ed., University Park Press, Baltimore, 1972, pp. 241-296.

46. W. Gordy and M. Kessler, Phys. Rev., **12**, 644 (1947); W. Gordy, Pure Appl. Phys., **11**, 403 (1965); P. Helminger, F. C. DeLucia and W. Gordy, Phys. Rev. Lett. **25**, 1397 (1970); P. Helminger, R. L. Cook and F. C. DeLucia, J. Molec. Spectrosc. **40**, 125 (1971).

47. J. Gilbert, Rev. Sci. Instr., **41**, 1050 (1970); T. Torring and E. Schnabel, Z. Physik, **204**, 198 (1967).

48. M. Winnewisser, Z. Angew. Physik, **30**, 359 (1971).

49. M. Lichenstein, J. J. Gallagher, and R. E. Cupp, Rev. Sci. Instr., **34**, 843 (1963).

50. E. P. Valkenburg and V. E. Derr, Proc. IEEE, **54**, 493 (1966).

51. W. D. Gwinn, A. C. Luntz, C. H. Sederholm, and R. Millikan, J. Comput. Phys., **2**, 439, (1968).

52. R. C. Woods and T. A. Dixon, Rev. Sci. Instr., **45**, 1122 (1974).

53. B. Haase, R. Gegenheimer, K. H. Peting, and W. Zeile, Processor Controlled Microwave Spectrometer for Trace Detection, N-73-32037 (NTIS Accession No.) (1973).

REFERENCES

54. R. Kewley, K. V. L. N. Sastry, M. Winnewisser, and W. Gordy, J. Chem. Phys., **39**, 2856 (1963), R. Kewley, K. V. L. N. Sastry, and M. Winnewisser, J. Molec. Spectrosc., **10**, 418 (1968).

55. L. Hrubesh, R. Anderson, and E. Rinehart, Rev. Sci. Instr., **42**, 789 (1971); L. Hrubesh, Radio Sci., **8**, 167 (1973).

56. L. Hrubesh, Radio Sci., **8**, 167 (1973), Rev. Sci. Instr., **41**, 595 (1970).

57. J. Hoeft, F. J. Lovas, F. Tiemann, and T. Törring, Z. Angew Physik, **31**, 5, 337 (1971); 6 (1971).

58. G. E. Jones and E. T. Beers, Anal. Chem., **43**, 656 (1971).

59. W. F. White, Qualitative Analysis of Gas Mixtures by Microwave Spectroscopy, presented at 155th National Meeting of American Chemical Society, San Francisco, California, March 31/April 5, 1968; Applied Microwave (MRR) Spectroscopy Problems and Progress at 162nd National Meeting of American Chemical Society, Washington, D. C., September 13-17, 1971.

60. Jean-Jacques Masini, A. Bouchy, and G. Roussy, J. Chemie Physique, **73**, 97 (1976).

61. T. Kondo, K. Hirota, and Y. Morino, J. Molec. Spectrosc., **28**, 471 (1968).

62. U. V. Reichert and H. Hartman, Z. Naturforsch, **27a**, 983, 989 (1972).

63. J. W. Bevan, A. C. Legon, D. J. Millen, and S. C. Rogers, J. C. S. Chem. Commun., 341 (1975).

64. W. Steinbach and W. Gordy, Phys. Rev. A, **11**, 729 (1975).

65. H. Hiroshe Irie, J. Fukudome, T. Ohmori, and J. Tanaka, J. C. S. Chem . Commun., (1975).

66. H. W. Kroto, B. M. Landsberg, R. J. Suffolk, and A. Vodden, Chem. Phys. Lett., **29**, 265 (1974).

REFERENCES

67. R. D. Suenram and D. R. Johnson, J. Molec. Spectrosc., 65, 239 (1977).

68. M. Winnewisser and J. J. Christiansen, Chem. Phys. Lett., 37, 270 (1976).

69. J. Høg and T. Pedersen, J. Molec. Spectrosc., 61, 243 (1976).

70. P. S. Bryan and R. L. Kuczkowski, Inorg. Chem., 10, 200 (1971).

71. P. S. Bryan and R. L. Kuczkowski, Inorg. Chem., 11, 553 (1972).

72. R. J. Skyally, P. G. Szanto, T. G. Anderson, and R. C. Woods, Astrophys. J., 204, L143 (1976).

73. A. C. Legon, D. J. Miller, and P. J. Mjöberg, Chem. Phys. Lett., 47, 589 (1977).

74. S. Saito, J. Molec. Spectrosc., 65, 229 (1977).

75. T. M. Sanders, A. L. Schawlow, G. C. Dousmanis, and C. H. Townes, Phys. Rev., 89, 1158 (1953).

76. F. X. Powell and D. R. Lide, J. Chem. Phys., 41, 1413 (1964), Shuji Saito, Helv. Chemica Acta, 54, 273 (1971).

77. S. Saito, J. Chem. Phys., 53, 2544 (1970).

78. D. J. Meschi and R. J. Myers, J. Molec. Spectrosc. 3, 405 (1959).

79. F. J. Lovas, E. Tiemann, and D. R. Johnson, J. Chem. Phys., 60, 5005 (1974).

80. T. Amano and E. Hirota, J. Molec. Spectrosc., in press (1978).

81. D. R. Johnson, F. X. Powell, Sci., 164, 950 (1969).

82. T. Amano, S. Saito, E. Hirota, and Y. Morino, J. Molec. Spectrosc., 32, 97 (1969).

83. T. Amano, E. Hirota, and Y. Morina, J. Molec.

REFERENCES

Spectrosc., 27, (1968); T. Amano, S. Saito, E. Hirota, Y. Morino, D. R. Johnson, and F. X. Powell, J. Molec. Spectrosc., 30, 275 (1969).

84. F. X. Powell and D. R. Johnson, J. Chem. Phys., 50, 4596 (1969).

85. S. Saito, in preparation.

86. F. J. Lovas and D. R. Johnson, J. Chem. Phys., 55, 41 (1971).

87. S. Saito and T. Amano, J. Molec. Spectrosc., 34, 383 (1970).

88. S. Saito and K. Takagi, Astrophys. J., 175, L47 (1972).

89. S. Saito, Astrophys. J., 178, L53 (1972).

90. V. M. Rao, R. F. Curl, Jr., R. L. Timms, and J. L. Margrave, J. Chem. Phys., 43, 2557 (1965).

91. H. Takeo, R. F. Curl, Jr., and P. W. Wilson, J. Molec. Spectrosc., 38, 464 (1971).

92. W. E. Steinmetz, J. Am. Chem. Soc., 96, 685 (1974).

93. M. S. Farag and R. K. Bohn, J. Chem. Phys., 62, 3946 (1975).

94. T. Ikeda, R. Kewley, and R. F. Curl, Jr., J. Molec. Spectrosc., 44, 459 (1972).

95. The spectra recorded on HP Model 864A microwave rotational spectrometer were obtained form Dr. L. H. Scharpen of the Hewlett-Packard Co., Palo Alto, California in 1975.

96. N.S. True and R. K. Bohn, J. Chem. Phys., 62, 3951 (1975).

97. L. W. Hrubesh, E. A. Rinehart, and R. E. Anderson, J. Molec. Spectrosc., 36, 354 (1970).

98. W. V. Smith and R. Howard, Physc. Rev., 79, 128 (1950); R. H. Hughes, Ann. N.Y. Acad. Sci., 55, 872 (1952).

REFERENCES

99. C. H. Townes, A. N. Holden, and F. R. Merritt, Phys. Rev., 71, 64 (1947).

100. D. H. Baird and G. R. Bird, Rev. Sci, Instr., 25, 319 (1954).

101. A. S. Esbitt and E. B. Wilson, Jr., Rev. Sci. Instr., 34, 901 (1963).

102. D. R. Lide, Jr. and D. Christensen, J. Chem. Phys., 35, 1374 (1961).

103. Y. Sakurai, Y. Kaneda, T. Kondo, K. Hirota, T. Onishi, and K. Tamaru, Trans. Faraday Soc., 67, 3094 (1971).

104. T. Kondo, S. Saito, and K. Tamaru, J. Am. Chem. Soc., 96, 6857 (1974).

105. G. F. Crable, Study of Air Pollutants by Microwave Spectroscopy, Report NTIS PB212-554 (1972); G. F. Crable and J. C. Wahr, J. Chem. Phys., 51, 181 (1969).

106. R. F. Curl, T. Ikeda, R. S. Williams, S. Leavell, and L. H. Scharpen, J. Am. Chem. Soc., 95, 6182 (1973).

107. R. Varma and R. F. Curl, J. Phys. Chem., 80, 402 (1976).

108. E. Saegebarth and E. B. Wilson, Jr., J. Chem. Phys., 46, 3088 (1967).

109. R. F. Curl, Jr., J. Molec. Spectrosc., 29, 375 (1969).

110. E. A. Rinehart, NASA CR-107171 (1969).

111. W. F. White, private communication (1975).

112. Y. Morino and E. Hirota, J. Chem. Soc. Jap., 85, 535 (1964).

113. S. Saito, Tetrahedron Lett., 48, 4961 (1968).

REFERENCES

114. S. Saito, Bull. Chem. Soc. Jap., __42__, 667 (1969).

115. S. Saito and C. Wentrup, Helv. Chim. Acta, __54__, 273 (1971).

116. Y. Saburai, Y. Kaneda, S. Kondo, E. Hirota, T. Onishi, and K. Tamaru, Bull. Chem. Soc. Jap., __41__, 1496 (1968).

117. Y. Sakurai, T. Onishi, and K. Tamaru, Bull. Chem. Soc. Jap., __45__, 980 (1972).

118. T. Ueda and K. Hirota, J. Phys. Chem., __74__, 4216 (1970).

119. T. Kondo, M. Ichikawa, S. Saito, and K. Tamaru, J. Phys. Chem., __77__, 299 (1973).

120. Y. Morino, J. Molec. Spectrosc., __19__, 1 (1973).

121. S. Naito, T. Kondo, M. Ichikawa, and K. Tamaru, J. Phys. Chem., __76__, 2184 (1972).

122. A. L. Dent and R. J. Kokes, J. Am. Chem. Soc., __92__, 6709 (1970).

123. T. Kondo, M. Ichikawa, S. Saito, and K. Tamaru, Bull. Chem. Soc. Jap., __45__, 1580 (1972).

124. S. Naito, M. Ichikawa, S. Saito, and K. Tamaru, J. Chem. Soc., Faraday I, __69__, 685 (1973).

125. K. Hirota and Y. Hironaka, Tetrahedron Lett., 25, 1645 (1964).

126. Y. Hironaka and K. Hirota, Tetrahedron Lett., 22, 2437 (1966).

127. H. Kim and W. D. Gwinn, Tetrahedron Lett., 37, 2535 (1964).

128. F. J. Lovas and D. R. Lide, Jr., Advances in High Temperature Chemistry, Vol. 3, Academic Press, New York 1971.

REFERENCES

129. D. R. Johnson and F. X. Powell, Science, <u>169</u>, 679 (1970).

130. D. R. Johnson and F. J. Lovas, Chem. Phys. Lett., <u>15</u>, 65 (1972).

131. F. J. Lovas, F. O. Clark, and E. Tiemann, J. Chem. Phys., <u>62</u>, 1925 (1975).

132. C. W. Gillies and R. L. Kuczkowski, J. Am. Chem. Soc., <u>94</u>, 6337 (1972); <u>94</u>, 7609 (1972).

133. R. P. Lattimer, C. W. Gillies, and R. L. Kuczkowski, J. Am. Chem. Soc., <u>95</u>, 1348 (1973).

134. F. J. Lovas and R. D. Suenram, unpublished internal report, National Bureau of Standards, Washington, D.C. (1977).

135. G. E. Jones and R. L. Cook, CRC Crit. Rev. in Anal. Chem., <u>3</u>, 455 (1974).

136. C. H. Townes, A. N. Holden, and F. R. Merritt, Phys. Rev., <u>71</u>, 64 (1947).

137. C. S. Wu, C. H. Townes, and L. Feldman, Phys. Rev., <u>76</u>, 692 (1949).

138. A. L. Southern, H. W. Morgan, G. W. Kielholtz, and W. V. Smith, Anal. Chem., <u>23</u>, 1000 (1952).

139. J. Weber and K. J. Laidler, J. Chem. Phys., <u>19</u>, 1089 (1951).

140. Hewlett Packard Application Note No. 840-4.

141. Hewlett Packard Application Note No. 840-5.

142. J. T. Funkhouser, S. Armstrong, and H. W. Harrington, Anal. Chem., <u>40</u>, 22A (1968).

143. W. F. White, Chem. Eng. Progr., <u>63</u>, 62 (1966).

144. R. W. Graff, Department of Electrical Engineering, Drexel University, personal communication.

REFERENCES

145. H. T. Buscher, B. Leskover, and W. F. Kolbe, University of California, Lawrence Livermore Laboratory, Report 4469, February 16, 1976.

146. L. W. Hrubesh, A. S. Maddux, D. C. Johnson, and J. N. Nielsen, University of California, Lawrence Livermore Laboratory, Report UCID-16488, February 1974.

147. L. W. Hrubesh, A. S. Maddux, and J. N. Nielsen, paper No. #14 presented at Pacific Analytical Conference, November 1975.

148. W. F. White, NASA Langley Research Center, Langley, VA, personal communication, April 1973.

149. E. A. Rinehart, Physics Department University of Wyoming, Laramie, Wyoming, personal communication, January 1973.

150. E. A. Rinehart and J. J. Fletcher, Determination of Sulfur Dioxide in Polluted Air by MRR Spectroscopy, presented at 167th ACS Meeting, Los Angeles, Calif. (1974).

151. D. N. Kaye, Electron. Design, 10, 50 (1971).

152. T. H. Maugh, II, Science, 177, 685 (1972).

153. J. A. Hodgeson, W. A. McClenny, and P. L. Hanst, Science, 182, 248 (1973).

154. R. L. Morrison, A. S. Maddux, and L. W. Hrubesh, University of California, Lawrence Livermore Laboratory Report UCRL-51945 (1975).

155. T. M. Srinivasan, Physica Scripta, 7, 84 (1973).

156. R. L. Morrison, University of California, Lawrence Livermore Laboratory, unpublished work.

157. A. N. Aleksandrov and G. J. Tysovskii, Zh. Anal. Khim., 22, 1286 (1967).

158. E. A. Rinehart, Anal. Chem., 49, 249A (1977).

REFERENCES

159. W. C. Easley, NASA Tech. Brief Langley Research Center, No. B73-10228, August 1973.

160. H. Uehara and Y. Ijuuin, Chem. Phys. Lett., $\underline{28}$, 597 (1974); M. Tanimoto and H. Uehara, Environ. Sci. Technol., $\underline{9}$, 153 (1975).

161. H. W. Harrington, Hewlett Packard Co., Scientific Instruments Division, Palo Alto, CA, personal communication, May 1972.

162. E. A. Rinehart, Physics Department, University of Wyoming, Laramie, August 1972.

163. C. A. Harris and G. E. Jones, Mississippi Academy of Science Meeting, University of Mississippi, Mississippi City, Mississippi, 1971.

164. F. W. Karasek, Res. Dev. Mag., December 1972, p. 30.

165. Dimethyl Silicone Membranes are supplied by General Electric Medical Products Division, Schenectady, New York.

166. L. W. Hrubesh, A. S. Maddux, Jr., and D. C. Johnson, paper No. 82 presented at 25th Pittsburg Conference on Analytical Chemistry, March 3, 1974.

167. M. J. Lazarus, S. Novak, and E. D. Bullimore, Microwave J., $\underline{14}$, 43 (1971).

168. F. J. Lovas, ISA Trans., $\underline{14}$, 145 (1975).

169. L. W. Hrubesh and J. C. Kantor, paper No. 191 presented at Pittsburg Conference on Analytical Chemistry, March 10, 1976.

170. J. C. Kantor and L. W. Hrubesh, University of California, Lawrence Livermore Laboratory, Report UCID-17027, October 1976.

REFERENCES

171. L. W. Hrubesh, A. S. Maddux, Jr., and D. C. Johnson, University of California, Lawrence Livermore Laboratory Report UCID-17513, June 1975.

172. R. L. Morrison, University of California, Lawrence Livermore Laboratory, unpublished work.

AUTHOR INDEX

Aleksandrov, A. N., 151
Anderson, R., 57
Armstrong, S., 136, 138

Baird, D. H., 92
Beaudet, R. A., 13
Beers, E. T., 63, 64
Bird, G. R., 92

Chung, C. S. C., 13
Clark, F. O., 128
Coles, D. K., 27
Cook, R. L., 134
Crable, G. F., 100-103, 115, 119, 146, 147
Curl, R. F., 80, 107, 112, 117, 122

Delfino, A. B., 13
Dixon, T. A., 44

Easley, W. C., 152
Esbitt, A. S., 92, 122, 135

Fletcher, J. J., 149
Funkhouser, J. T., 136, 138

Gaumann, T., 13
Gegenheimer, R., 44, 45, 50
Gwinn, W. D., 44, 45, 47, 130

Haase, B., 44, 45, 50
Harrington, H. W., 33, 113, 115, 119, 152
Harris, C. A., 152
Hironaka, Y., 129
Hirota, E., 126, 129
Hoeft, J., 58
Holden, A. N., 92, 133
Hrubesh, L., 55, 56, 57
Hughes, H. R., 18

Ijuuin, Y., 152
Ikeda, T., 80, 112

Jones, G. E., 63, 64, 134, 152

Kaneda, Y., 95
Karplus, B., 25
Kewley, R., 51
Kim, H., 130
Kirchoff, W. H., 14
Kivelson, D., 14
Kneubuhl, F., 13
Kondo, T., 95

Laidler, K. J., 136
Lazarus, M. J., 158
Lovas, F. J., 58, 128, 132, 161
Luntz, A. C., 44, 45, 47

McAfee, K. B., 18
Morgan, H. W., 133
Morino, Y., 126
Morrison, R. L., 151, 181, 182

Novak, S., 158

Pollnow, G. F., 13

Ramaprasad, K. P., 13
Rasukolb, R. F., 39, 92, 93, 95, 122
Rinehart, 52, 118, 122, 149, 151, 152

Saito, S., 95, 130, 131
Saegebarth, E., 107
Sakurai, Y., 95
Sastri, K. V. L. W., 51
Scharpen, L. H., 39, 92, 93, 95, 122
Srinvasan, T. M., 150
Southern, A. L., 133
Suenram, R. D., 132

Townes, C. H., 26, 92, 133
Tysovskii, G. J., 151

Uehara, H., 152

Van Vleck, J. H., 27
Varma, R., 107

Watson, J. K. G., 14
Weber, J., 136
Weisskopf, V. F., 27
Wentrup, C., 131
White, W. H., 39, 44, 45, 49, 50, 64, 75, 76, 77, 119, 137, 138, 139, 159, 174

Wilson, E. B., Jr., 14, 18, 21, 42, 92, 107, 122, 135
Winnewisser, G., 45
Winnewisser, M., 42, 44, 45, 51
Woods, R. C., 44

Zeile, W., 45

SUBJECT INDEX

Absolute intensity, 27, 28
  measurement, 106-114
Absorption cell, 19, 20, 37
  free space cells, 42, 51, 52
  parallel plate absorption cell, 53
  resonant cavity absorption cell, 53
Absorption coefficient, 27-34
  integrated absorption coefficient, 30, 91
    measurement, 106
  Γ-intensity coefficient, 32-34
    measurement, 114-120
  peak intensity coefficient, 29, 30
    measurement, 92-99, 133-135
Abundance of isotopes, 92, 133-135
Air pollution measurement and/or monitoring, 61, 145-150
  cigarette smoke analysis, 152, 153
  engine exhaust analysis, 151, 152
  portable gas monitors, 170-180
    ammonia monitor, 147, 174, 176-177
    formaldehyde monitor, 73, 147, 170-174
    multigas meter, 179
    permeable membranes, 147, 157, 162

Backward wave oscillator, 18-19, 35, 37
Bridge method for intensity measurement on HP Model 8460A Spectrometer, 38, 39, 111, 112

Cavity spectrometer, 53, 54, 55
  analysis of ammonia in blood, 150
  Gunn diode effect oscillator, 56
  Gunn diode resonant-cavity spectrometer, 55-56
  portable gas monitors, 170-177
Centrifrugal distortion effects, 14
Collision related line broadening, 24-26, 88-90
Commercial spectrometers, 35
  Cambridge Scientific Instruments (CSI) microwave rotational (MR) spectrometer, 39-41
    integrator device, 40
    frequency measurement, 21
    frequency range useful for chemical analysis, 65
  Hewlett Packard (HP) Model 8460A MR spectrometer, 35-39
    backward wave oscillator (BWO), 35-37
    low resolution microwave band spectroscopy (LMWBS), 77-81
    pin diode and calibration arm, 39, 99, 116-119
    Stark cell, 37
    Stark modulation, 21, 22
    wave guide bridge system, 38, 39
  intensity measurement, 92-99, 106, 114-120, 133-175
  resolution, 59, 60
  sample size, 123
  sensitivity, accuracy and precision, 61, 121-123
Computer interfaced Gunn diode MR cavity spectrometer, 56-58
Computer interfaced MR spectrometer, 44-50
  line strength, and frequency calculation, 4-18
  intensity and frequency, 4-18
  qualitative analysis, 74-77
    second derivative spectrum, 50
    S:N ratio improvement through time integration, computer filtering and digital averaging, 47-49
    specificity in MRS, 62-68
  quantitative analysis, on automated MR spectrometer, 92-95, 137-145, 149-152
    gas mixture analysis, 137-145
    gas trapped on chemsorb material, 149
    hydrogen-deuterium exchange reaction, 92-95, 126-129
    process control sensors, 159-169
    reference data, 74-75

SUBJECT INDEX

Detection and characterization of chemical species by MRS, 68-74
   resolution and sensitivity of MRS, 59-62
   specificity of MRS, 62-68
   stable molecules, reaction intermediates and free radicals, 70-74
Doppler effect in broadening rotational lines, 24-25

Energy differences of rotomers, between reactants and products of chemical reaction, 107-112
Engine exhaust analysis, 151-152

Fabry-Perot millimeter wave spectrometer, 43
Formaldehyde monitors, 73, 147, 170-174
Free energy function determination for pure substance, 112, 113
Free radical identification and characterization, 73
Free space MR absorption cell, 42, 51, 52
Frequencies and intensities of rotational transitions, 4-18
   centrifugal distortion effects, 14
   computer programs, 13
   internal rotation, 15
   nuclear quadrupole effects, 17
   rotational energy levels, 9, 10, 12
   selection rules, 13
Frequency measurement, 21
Frequency range for chemical analysis, 65

Gunn effect diode cavity MR spectrometer, 56-58, 154-159
   Gunn effect diode detector, 158
   Gunn effect diode oscillator, 56

Half width of line at half height (HWHH), 100-104, 105-106
   Gaussian line shape, 24, 25
      Doppler broadening, 24
      wall collision related broadening, 25
   Lorentzian line shape, 26
      intermolecular collision related broadening, 88-91
Hyperfine splitting of MR spectra due to nuclear electronic quadrupole effect, 17

Identification and characterization by MRS, 68-74
   components in gas mixture, 135-145
   computer interfaced MR spectrometer, 44-50, 74-77
   computer programs for calculating MR frequencies, 13
   large organic molecules by LMWBS, 82-85
   line strength and intensity from reference data, 74-75
   reaction products, 69-74, 92-99, 126-129, 152-154, 182-185
   resolution and sensitivities of MRS, 59-62
   specificity of MRS, 62-68
   useful frequency range for chemical analysis, 65
Inexpensive analytical spectrometers, 154-159
   Gunn effect-diodes, 156-157
   UCLL spectrometers, 158-159
Intensity of rotational transitions, 23-27
   absorption coefficient α_ν, 27-29
   Γ-intensity coefficient under conditions of power saturation, 114-120
   line shape, 24-26
   measurement, of integrated intensities, 106
      of peak intensities, 92-100
      of peak intensity and HWHH, 87-91
      of relative and absolute intensities, 100-114
         bridge method, 104, 105
         peak height and HWHH, 105, 106
   population of rotational states, 4, 5
      absorption coefficient α_ν, 27, 29
      Γ-intensity coefficient, 32, 34
      integrated absorption coefficient, 30, 91
      peak absorption coefficient, 29, 30
      power saturation, 27, 32
   power saturation, 112-120

Line strength and intensity reference MRS data, 74-75
Line width (HWHH), 100-104, 105-106
Low resolution microwave band spectroscopy (LMWBS), 77-81
   analysis of large organic molecules, 82-85

Membrane separators, 147, 148, 156

Microwave spectrometer, 18-22
  combined GC-microwave analytical spectrometer, 180-185
  commercial MR spectrometers, 35-41
    CSI MR spectrometer, 39-41
    HP Model 8460A, 35-39
  computer interfaced MR spectrometer, gas mixture analysis, qualitative, 74-76
    gas mixture analysis, 135-145
    routine analytical spectrometer at UCLL, 140-145
  frequency range, 2
  Gunn diode microwave cavity spectrometer, 56-58
  Gunn diodes, 55, 56
  multigas monitors, 178-180
  portable gas monitors, 170-176
    ammonia monitor, 174-178
    formaldehyde monitor, 170-174, 175
  resolution, sensitivity, accuracy and precision, 59-60, 121-123
  useful frequency range for chemical analysis, 65
Millimeter wave spectrometers, 42-44, 51-52
  millimeter wave Fabry-Perot spectrometer, 43-44
  millimeter wave spectrometer with range up to 250 GC, for detection of free radicals, 51-52
  video-type millimeter wave spectrometer, 51-52

NBS microwave spectral tables, 28, 63, 75
Nuclear electric quadrupole effect, 17

"Oil plant-flares" gas analysis, 149
Oil shale cracking gas analysis, 182-185
Ozonolysis reaction product analysis, 130-133

Peak intensity of rotation transitions, 87-91
  techniques, of measurement, 92-100
  of relative and absolute intensity measurements, 100-114
Population of rotational states, 4, 5
Power saturation, 26, 27, 32-34

Qualitative analysis, 59
  application of MRS for identification and characterization, 68-74
    molecular species, 70-72
    transient species and free radicals, 73, 74
  computation of frequency of rotational transitions, 13
  on computer controlled MR spectrometer, 74-76
  line strength and intensity reference MR data, 74-75
  LMWBS for, 77-81
    applications in identification and characterization of large organic molecules, 82-85
  resolution and sensitivity of MRS, 59-61
  specificity of high resolution MRS for, 61-64
Quantitative analysis, under conditions of low power level, 86-87
  air pollution measurements, 145-150
  portable gas monitors, 170-180
  cigarette smoke analysis, 152
  engine exhaust analysis, 151-152
  integrated intensity, 91
  measurement, 106
  peak intensity of rotational transitions, 87-91
  isotope ratio measurement, 133-135
  line broadening, 88-91
  measurement of peak intensity, 92-100
  process control sensors, 159-170
  relative and absolute intensity measurement, 100-113
  mixture analysis by automated MRS, 135-145
  under conditions of power saturation, 86-87
  Γ-intensity coefficient, 32-34
    measurement for chemical analysis, 114-120

Reaction study, 126-130
  hydrogen-deuterium exchange, 92-97, 126-129
  mechanism of chemical reactions, 126-128, 129, 130
  ozonolysis reaction, 130-133

pyrolysis reaction, 128, 130-133

Sample size for MRS, 123
Sensitivity, accuracy and precision of MRS, 121-123
   and resolution of MR spectra, 16, 62, 121-123
Shape of rotational transitions, 23-27
Stark modulation, 20, 22

Thermodynamic function determination, 107-114
   energy differences of chemical species in ground state, 107-112
   free energy functions of pure substances, 112-114

Uncertainty in chemical analysis by MRS, 120-124